Conformal Field Theory
and Topology

Translations of

MATHEMATICAL MONOGRAPHS

Volume 210

Conformal Field Theory and Topology

Toshitake Kohno

IWANAMI SERIES IN MODERN MATHEMATICS

American Mathematical Society
Providence, Rhode Island

場の理論とトポロジー

FIELD THEORY AND TOPOLOGY

by Toshitake Kohno

Copyright © 1998 by Toshitake Kohno

Originally published in Japanese
by Iwanami Shoten, Publishers, Tokyo, 1998

Translated from the Japanese by the author.

2000 *Mathematics Subject Classification*. Primary 54C40, 14E20;
Secondary 46E25, 20C20.

Library of Congress Cataloging-in-Publication Data

Kohno, Toshitake.

[Ba no riron to toporoji. English]

Conformal field theory and topology / Toshitake Kohno ; [translated from the Japanese by the author].

p. cm. — (Translations of mathematical monographs, ISSN 0065-9282 ; v. 210)

(Iwanami series in modern mathematics)

Includes bibliographical references and index.

ISBN 0-8218-2130-X (acid-free paper)

1. Conformal invariants. 2. Quantum field theory. 3. Topology. I. Title. II. Series. III. Series: Iwanami series in modern mathematics.

QC174.52.C66 K6413 2002

530.14′3—dc21

2002018248

10 9 8 7 6 5 4 3 2 1   07 06 05 04 03 02

Contents

Preface to the English Edition

This is a translation of my book originally published in Japanese by Iwanami Shoten, Publishers. The aim of this book is to provide the reader with a concise introduction to geometric aspects of conformal field theory and its application to topological invariants. In the present English edition, I added more details to the original Japanese edition.

I would like to thank the American Mathematical Society for publishing this English edition and their staff for excellent support.

Toshitake Kohno

July, 2001

Preface

Geometry and physics have been developed with a strong influence on each other. One of the most remarkable interactions between geometry and physics since 1980 is an application of quantum field theory to topology and differential geometry. This book will focus on a relationship between two dimensional quantum field theory and three dimensional topology, which has been studied intensively since the discovery of the Jones polynomial in the middle of the 1980's and Witten's invariant for 3-manifolds derived from Chern-Simons gauge theory. An essential difficulty in quantum field theory comes from the infinite dimensional freedom of a system. Techniques dealing with such infinite dimensional objects developed in the framework of quantum field theory have been influential in geometry as well. The aim of this book is to give a treatment for a rigorous construction of topological invariants originally defined as partition functions of fields on manifolds.

In quantum field theory, we often encounter several completely different expressions for the same object, which are expected to be equivalent. For example, in many situations, quantization by Feynman's path integral and the geometric quantization based on complex line bundles are believed to be equivalent. We sometimes find new relations between geometric objects when they are translated into the language of quantum field theory. One of the main themes of this book is a connection between Chern-Simons gauge theory and conformal field theory, which has been obtained by investigating quantum Hilbert space for the partition function of the Chern-Simons functional in two ways. Although such a method is suggestive and sheds new light on many aspects in geometry, we have to keep in mind that equivalence in physics might not be a rigorously established fact in mathematics. To show the equivalence we will need to develop extra tools in geometry, which provides interesting problems in geometry.

The book is organized in the following way. In the Introduction we start from classical mechanics and explain basic background materials in quantum field theory and geometry. In Chapter 1, we deal with conformal field theory based on geometry of loop groups. First, we recall complex line bundles on loop groups, affine Lie algebras and the Virasoro Lie algebra. Then, we will formulate the space of conformal blocks and the KZ equation, which are central objects in conformal field theory. Usually, the KZ equation is derived from the operator product expansion of field operators, but in this book we will emphasize geometric points of view. The main object of Chapter 2 is the holonomy of conformal field theory. By means of linear representations of braid groups and mapping class groups appearing in conformal field theory, we will define the Jones polynomial for links and Witten's invariants for 3-manifolds. We will show that this provides an example of topological quantum field theory in the sense of Atiyah. In Chapter 3, we treat Chern-Simons perturbation theory. As the perturbative expansions of the Jones polynomial we obtain a series of finite type invariants for knots, which will be treated systematically as Vassiliev invariants. As a universal expression for such invariants we will describe the Kontsevich integral. We will stress how Chen's iterated integrals for braids are generalized to the Kontsevich integral for knots. Finally, we discuss topological invariants for 3-manifolds derived from Chern-Simons perturbation theory.

I have tried to make the sections of the book fairly independent, even at the cost of some repetition. The contents of this book cover only a small portion of interactions between quantum field theory and topology, but I hope that the reader might feel that these interactions are fruitful.

Toshitake Kohno

July, 1998

Introduction

We begin by recalling basic mathematical structures of classical mechanics, quantum mechanics and quantum field theory. We then proceed to explain a general framework to obtain topological invariants from quantum field theory.

Lagrangians and Hamiltonians in classical mechanics

Let N be an n-dimensional smooth manifold and denote by TN its tangent bundle. Suppose that there is a smooth function $\mathcal{L}(t, x, \xi)$ called a *Lagrangian* defined over TN depending on a time parameter t. Here $x = (x_1, \cdots, x_n)$ is a local coordinate for N and $\xi = (\xi_1, \cdots, \xi_n)$ is a coordinate for the tangent space. Let $\gamma : [a, b] \to N$ be a smooth curve on N. The integral

$$S = \int_a^b \mathcal{L}(t, \gamma(t), \gamma'(t)) dt$$

is called the *action integral*. We fix the endpoints $\gamma(a)$ and $\gamma(b)$ and try to minimize the above action integral. It turns out that a curve γ corresponding to a critical point of the action integral satisfies the *Euler-Lagrange equations*

$$\frac{d}{dt}\frac{\partial \mathcal{L}}{\partial \xi_j} = \frac{\partial \mathcal{L}}{\partial x_j}, \quad j = 1, \cdots, n.$$

Let us consider the case when N is an n dimensional Euclidean space. To recover Newton's equations for a particle we take

$$\mathcal{L} = \frac{1}{2}m \sum_{j=1}^n \xi_j^2 - V$$

where V is a potential. A great advantage of the Lagrangian formulation is that it reveals a connection between symmetries of a physical system and its conservation laws.

To explain the Hamiltonian formulation, first let M be a smooth manifold of dimension $2n$. A *symplectic form* ω is a non-degenerate, closed 2-form on M. Here we say that ω is non-degenerate if and only if $\omega^n \neq 0$ at any point on M. A smooth manifold equipped with a symplectic form is called a *symplectic manifold*. Given a symplectic form ω, we have a one-to-one correspondence between 1-forms on M and smooth vector fields on M in the following way. Let X be a smooth vector field on M. We define a 1-form $\varphi = i(X)\omega$ by

$$\varphi(Y) = \omega(X, Y)$$

for a smooth vector field Y on M. The above correspondence

$$X \mapsto i(X)\omega$$

gives an isomorphism between the vector space of 1-forms on M and the vector space of smooth vector fields on M. Let f be a smooth function on M. We define a vector field X_f by

$$i(X_f)\omega = df.$$

We call X_f the *Hamiltonian vector field* for f. For smooth functions f and g on M we define the *Poisson bracket* $\{f, g\}$ by

$$\{f, g\} = -\omega(X_f, X_g).$$

The Poisson bracket is bilinear and anti-symmetric. We have the Jacobi identity

$$\{f, \{g, h\}\} + \{g, \{h, f\}\} + \{h, \{f, g\}\} = 0$$

and the relation

$$\{f, gh\} = \{f, g\}h + g\{f, h\}.$$

The second relation signifies that the operator $\{f, \ \}$ acts as a derivation on the space of smooth functions $C^\infty(M)$ on M. It follows from the definition that we have

$$\omega(X_f, X_g) = i(X_f)\omega(X_g) = df(X_g) = X_g f,$$

which is equal to $-X_f g$. We denote by $\mathcal{X}(M)$ the space of smooth vector fields on M. By means of the bracket $[X, Y]$ defined by

$$[X, Y]f = X(Yf) - Y(Xf), \ f \in C^\infty(M),$$

for $X, Y \in \mathcal{X}(M)$, the space of vector fields $\mathcal{X}(M)$ is equipped with a Lie algebra structure. The space of smooth functions $C^\infty(M)$ is a Lie algebra by the Poisson bracket. We have $[X_f, X_g] = X_{\{f,g\}}$, therefore, the correspondence

$$f \mapsto X_f$$

defines a Lie algebra homomorphism.

Given a smooth function H on M called a *Hamiltonian*, consider the associated Hamiltonian vector field X_H. On an integral curve of X_H, a smooth function f on M satisfies

$$\frac{df}{dt} = X_H f$$

where t is a parameter for the integral curve. Hence we have

$$\frac{df}{dt} = \{H, f\}.$$

In particular, the Hamiltonian H is constant on the above integral curve since

$$\frac{dH}{dt} = \{H, H\} = 0.$$

Complex line bundles and the quantization

Let M be a smooth manifold and E a smooth vector bundle on M. We denote by $T^*M_{\mathbf{C}}$ the complexification of the cotangent bundle T^*M of M. We write $\Gamma(E)$ and $\Gamma(T^*M_{\mathbf{C}} \otimes E)$ for the smooth sections of E and $T^*M_{\mathbf{C}} \otimes E$ respectively. A *connection* on E is a $\mathbf{C}$ linear map

$$\nabla : \Gamma(E) \to \Gamma(T^*M_{\mathbf{C}} \otimes E)$$

such that the Leibniz rule

$$\nabla(fs) = df \otimes s + f\nabla(s)$$

holds for $f \in C^\infty(M)$ and $s \in \Gamma(E)$. For a vector field $X \in \Gamma(T^*M_{\mathbf{C}})$ we define a linear map $\nabla_X : \Gamma(E) \to \Gamma(E)$ by

$$(\nabla_X s)(x) = (\nabla s)(X_x).$$

We call ∇_X the *covariant derivative* for X.

In the following, we deal with a *complex line bundle*, i.e., a complex vector bundle of rank 1. Let L be a complex line bundle with a Hermitian metric over a smooth manifold M. There is an open covering $M = \bigcup_j U_j$ such that L is trivialized as $U_j \times \mathbf{C}$ over U_j. Let

∇ be a connection on L. On each U_j the connection ∇ is expressed as

$$\nabla = d - 2\pi\sqrt{-1}\alpha_j$$

with a 1-form α_j locally defined on U_j. We see that $d\alpha_j$ defines a global 2-form on M, which is called the *first Chern form* of ∇. We denote it by $c_1(\nabla)$. The de Rham cohomology class $[c_1(\nabla)] \in H^2(M,\mathbf{R})$ does not depend on the choice of a connection ∇ and is called the *first Chern class* of L. The first Chern class $c_1(L)$ is contained in the image of the natural map

$$i : H^2(M,\mathbf{Z}) \to H^2(M,\mathbf{R}).$$

Conversely, given a closed 2-form ω on M such that the de Rham cohomology class of ω is contained in the image of $i : H^2(M,\mathbf{Z}) \to H^2(M,\mathbf{R})$, one can construct a complex line bundle L on M with a Hermitian connection ∇ such that $c_1(\nabla) = L$. In Chapter 1, we give an explicit construction of such a complex line bundle in the case when M is simply connected.

As explained in the last section, the basic dynamic equation of Hamiltonian mechanics is expressed using the Hamiltonian H and the Poisson bracket by

$$\frac{df}{dt} = \{H, f\}.$$

A framework of quantum mechanics due to Dirac might be stated in the following way. We associate a linear operator $\tilde{f}$ to a function f and suppose that the correspondence $f \mapsto \tilde{f}$ is linear and that

$$\tilde{f}_1\tilde{f}_2 - \tilde{f}_2\tilde{f}_1 = -\sqrt{-1}\hbar\tilde{f}_3$$

holds when $\{f_1, f_2\} = f_3$. We set $\sqrt{-1}\tilde{f}/\hbar = \widehat{f}$. The operators $\widehat{f}$ form a non-commutative algebra and we define the Lie bracket of operators by

$$[\tilde{f}_1, \tilde{f}_2] = \tilde{f}_1\tilde{f}_2 - \tilde{f}_2\tilde{f}_1.$$

The fundamental equation in quantum mechanics is

$$\frac{d\widehat{f}}{dt} = [\widehat{H}, \widehat{f}],$$

which is called the *Heisenberg equation*.

We represent a classical mechanical system as a symplectic manifold (M,ω) and regard $C^\infty(M)$ as a Lie algebra by the Poisson bracket. The above framework of quantum mechanics is interpreted as a representation of the Lie algebra $C^\infty(M)$ as linear operators acting on a certain Hilbert space. This is a basic idea of *quantization*. Let us try to perform this program of quantization by means of a complex line bundle on M. We suppose that the symplectic manifold (M, ω) satisfies the condition that ω determines an integral cohomology class in $H^2(M, \mathbf{Z})$. Then there exists a complex line bundle L with a Hermitian connection ∇ such that the first Chern form $c_1(\nabla)$ is equal to ω. We denote by $\Gamma(M, L)$ the space of smooth sections of L and we introduce an inner product by

$$\langle s_1, s_2 \rangle = \int_M (s_1(x), s_2(x)) \frac{\omega^n}{n!}$$

where $(s_1(x), s_2(x))$ stands for the Hermitian metric for the line bundle L. We denote by $\mathcal{H}$ the space of L^2 sections of L with respect to the above inner product. The space $\mathcal{H}$ has a structure of a Hilbert space. Let us consider ∇_{X_f}, the covariant derivative for the Hamiltonian vector field X_f. For a smooth function f on M we define a linear operator $\widehat{f}$ acting on $\mathcal{H}$ by

$$\widehat{f}s = \nabla_{X_f} s - 2\pi\sqrt{-1}fs, \quad s \in \mathcal{H}.$$

By means of the equality

$$\nabla_X \nabla_Y - \nabla_Y \nabla_X = \nabla_{[X,Y]} - 2\pi\sqrt{-1}\omega(X, Y)$$

together with $\omega(X_f, X_g) = -\{f, g\}$, we can verify that the operators $\widehat{f}$, $f \in C^\infty(M)$ satisfy the relation

$$[\widehat{f}, \widehat{g}] = \widehat{\{f, g\}}.$$

In other words, the correspondence $f \mapsto \widehat{f}$ determines a representation of $C^\infty(M)$ on the Hilbert space $\mathcal{H}$ as a Lie algebra.

We consider the case $N = \mathbf{R}^n$. The cotangent bundle $M = T^*N$ has a structure of a symplectic manifold with

$$\omega = \sum_{j=1}^{n} dp_j \wedge dq_j$$

where $q_1, \cdots, q_n$ are coordinate functions for N and $p_1, \cdots, p_n$ are coordinate functions for the tangent space. We put $\theta = \sum_{j=1}^{n} p_j dq_j$ and define a connection on a trivial complex line bundle $L = M \times \mathbf{C}$ by

$$\nabla = d - \sqrt{-1}\hbar^{-1}\theta.$$

For a smooth section s of L, i.e., a smooth complex valued function on M, we set

$$\tilde{f}s = \sqrt{-1}\hbar\nabla_{X_f}s + fs.$$

The operators $\tilde{q}_j$ and $\tilde{p}_j$ associated with the coordinate functions q_j and p_j are written as

$$\tilde{p}_j = -\sqrt{-1}\hbar\frac{\partial}{\partial q_j}, \quad \tilde{q}_j = \sqrt{-1}\hbar\frac{\partial}{\partial p_j} + q_j.$$

Let $\mathcal{H}_0$ be the subspace of $\Gamma(M, L)$ consisting of L^2 sections depending only on $q_1, \cdots, q_n$. Then, for $s \in \mathcal{H}_0$ we have

$$\tilde{p}_j s = -\sqrt{-1}\hbar\frac{\partial}{\partial q_j}s, \quad \tilde{q}_j s = q_j s.$$

Thus, we recover the canonical quantization in quantum mechanics. In general the whole space of L^2 sections of L is too large a Hilbert space for our purpose. The following polarization provides us with a method to extract a suitable subspace.

Let (M, ω) be a symplectic manifold of dimension $2n$ and $TM_{\mathbf{C}}$ the complexified tangent bundle of M. A subbundle V_P of $TM_{\mathbf{C}}$ is *integrable* if for any sections X and Y of V_P, $[X, Y]$ is a section of V_P. We say that V_P is *Lagrangian* if for any point $x \in M$ the fibre $(V_P)_x$ over x has dimension n and the symplectic form ω is identically 0 on $(V_P)_x$. A Lagrangian subbundle V_P of $TM_{\mathbf{C}}$ is called a *polarization* if it is integrable. Given a polarization V_P, we put

$$\mathcal{H}_P = \{s \in \mathcal{H} \mid \nabla_X s = 0, \ X \in \Gamma(M, V_P)\}$$

and we take $\mathcal{H}_P$ as a quantum Hilbert space instead of $\mathcal{H}$. In the above example of canonical quantization the subspace $\mathcal{H}_0$ is obtained from the polarization by the kernel of the projection map

$$d\pi : TM_{\mathbf{C}} \to TN_{\mathbf{C}}.$$

Let (M, ω) be a Kähler manifold and we take $TM^{(0,1)}$, the subbundle of $TM_{\mathbf{C}}$ consisting of $(0, 1)$ vectors, as a polarization. This is called a *Kähler polarization*. The Hilbert space $\mathcal{H}_P$ obtained from the Kähler polarization is the space of holomorphic sections $H^0(M, L)$. As an important application let us mention briefly the *Borel-Weil theory*. Let G be a compact Lie group and T its maximal torus. The quotient space G/T is called the flag manifold and has a structure of a Kähler manifold. The Borel-Weil theory gives a realization of any irreducible representation of G on the space of holomorphic sections of $H^0(G/T, L)$. This construction might be considered to be a prototype of conformal field theory described in Chapter 1.

Sigma models

Let us review sigma models, which will provide a basic framework in classical field theory. We fix a Riemannian manifold M. Let X be a compact oriented Riemannian manifold of dimension $d + 1$. A smooth map

$$\phi : X \to M$$

is called a *field*. We consider the set of such smooth maps ϕ and we denote it by

$$C_X = \mathrm{Map}(X, M),$$

which is the set of fields in the sigma model. Here the manifold X is a model of the space-time of space dimension d. In general, there are several variants of the notion of fields. For example, in gauge theory, the space of connections on a principal fibre bundle over X plays a role of the set of fields. We consider a functional

$$S_X : C_X \to \mathbf{R}$$

called an action. In the above sigma model we put

$$S_X(\phi) = \int_X |d\phi|^2 \, dv$$

where dv is the volume form of the Riemannian manifold X. A field ϕ minimizing the action S_X is called a *harmonic map*.

The following properties of the action S_X are easily verified and play an important role in our theory.

1. Let $f : X' \to X$ be an isometry. Then, f induces a map $f^* : C_X \to C_{X'}$ and we have $S_{X'}(f^*\phi) = S_X(\phi)$ for $\phi \in C_X$.
2. We denote by $-X$ the manifold X with the reversed orientation. Then, we have $S_{-X}(\phi) = -S_X(\phi)$ for $\phi \in C_X$.
3. Let $X = X_1 \sqcup X_2$ be the disjoint union of manifolds X_1 and X_2. We denote by ϕ_i the restriction of ϕ on X_i, $i = 1, 2$. Then,

$$S_X(\phi) = S_{X_1}(\phi_1) + S_{X_2}(\phi_2)$$

holds.

4. Let Y be a codimension one submanifold of X such that $X = X_+ \cup X_-$, $\partial X_+ = Y$ and $\partial X_- = -Y$. Then, for $\phi_+ \in C_{X_+}$ and $\phi_- \in C_{X_-}$ with $\phi_+|Y = \phi_-|Y$, there exists $\phi \in C_X$ such that

$$S_X(\phi) = S_{X_+}(\phi_+) + S_{X_-}(\phi_-)$$

holds.

Quantization by Feynman's path integral

In classical mechanics the trajectory of a particle between two points in the configuration space is determined as a path minimizing the action integral S for the Lagrangian of the system. On the other hand, in quantum mechanics, we consider the contribution of the probability function $e^{\sqrt{-1}S(\gamma)/\hbar}$ for any path γ connecting the two points. This is formally written as *Feynman's path integral*

$$\int e^{\sqrt{-1}S(\gamma)/\hbar} d\mu(\gamma)$$

where $d\mu$ is a measure on the space of paths connecting the above two points. However, a rigorous formulation of such an infinite dimensional integral is out of our reach except in a few cases where the integral is reduced to the iterations of Gaussian integrals. In the case $\hbar \to 0$ the rapid oscillations of $e^{\sqrt{-1}S(\gamma)/\hbar}$ will tend to cancel large contributions to the integral, but this cancellation will not occur at the critical points of $S(\gamma)$. Therefore, main contributions to the integral come from the critical points of S, i.e., the classical trajectories of the particle. In this way classical theory is recovered from quantum theory when $\hbar \to 0$.

Let X be an oriented smooth manifold without boundary and try to perform a program of quantization by Feynman's path integral. We need a measure μ_X on C_X and to compute the partition function

$$Z(X) = \int_{C_X} e^{\sqrt{-1}S(\phi)/\hbar} d\mu_X(\phi).$$

Unfortunately, a rigorous formulation of the above program is not available in many cases. We will take a different point of view to construct $Z(X)$ in a rigorous way. A key point is to decompose X into two parts $X_+ \cup X_-$ as in the last section and to formulate a theory for a manifold with boundary. Let us continue a formal discussion. Let X be an oriented smooth manifold with boundary Y. For $\alpha \in C_Y$

we define $C_X(\alpha)$ to be the set of $\phi \in C_X$ such that $\phi|Y = \alpha$. Again we formally write

$$Z(X)_\alpha = \int_{C_X(\alpha)} e^{\sqrt{-1}S(\phi)/\hbar} d\mu_X(\phi)$$

and consider $Z(X)$ as a function on C_Y. Our strategy is to define Z_Y, a certain space of functions on C_Y that $Z(X)$ should live in, and to write down expected properties for $Z(X)$. One of the main objects of this book is to formulate Z_Y in the case of the Wess-Zumino-Witten model and the Chern-Simons theory as sections of complex line bundles. We shall call Z_Y the *quantum Hilbert space*.

Several properties for Z_Y and $Z(X)$ are expected from the properties of S_X given in the last section. We will list some of them as axioms.

1. (orientation axiom) We denote by $-Y$ the manifold Y with reversed orientation. Then we have

$$Z_{-Y} = Z_Y^*$$

where Z_Y^* is the dual vector space of Z_Y.

2. (multiplicativity) For a disjoint union $Y = Y_1 \sqcup Y_2$ we have

$$Z_{Y_1 \sqcup Y_2} = Z_{Y_1} \otimes Z_{Y_2}.$$

3. (gluing axiom) For $X = X_+ \cup X_-$ with $\partial X_+ = Y$ and $\partial X_- = -Y$ we have

$$Z(X) = \langle Z(X_+), Z(X_-) \rangle$$

where the right hand side stands for the natural pairing between Z_Y and Z_Y^*.

Wess-Zumino-Witten models and loop groups

In Chapter 1, we give a description of complex line bundles on loop groups and formulate the Wess-Zumino-Witten model. Based on this geometric construction we develop conformal field theory. Let Σ be a compact Riemann surface and $f : \Sigma \to G$ a smooth map, where G is a compact Lie group. In this book we deal with the case $G = SU(2)$. For f we will define the Wess-Zumino-Witten action $S_\Sigma(f)$ and try to compute its partition function, which was originally defined by Feynman's path integral of $\exp(-S_\Sigma(f))$ over the space of smooth maps $f : \Sigma \to G$. Our main object is to consider the case of a Riemann surface with boundary and to show that $\exp(-S_\Sigma(f))$ is considered to be an element of a fibre of a complex line bundle on the loop group. The partition function will be regarded as a section of this complex line bundle. The space of sections of the complex line bundle on the loop group admits an action of the affine Lie algebra, which is a central extension of the infinitesimal version of the loop group. It turns out that the partition function should be invariant under the action of holomorphic maps. This provides us with a constraint on the partition function as a section of the complex line bundle.

Motivated by the above consideration, we are going to pursue the following representation theoretical construction. We start from representations of the affine Lie algebra and we extract the space of coinvariant tensors which are invariant under the action of holomorphic maps. The resulting space, in which the partition function should live, will be called the space of conformal blocks. The space of conformal blocks is regarded as the space of operators with parameters on the Riemann surface which have invariance under conformal transformations. This is a framework of conformal field theory. In particular, in the case of the Riemann sphere, we shall describe explicitly the fusion rules, which give the dimension formula for the space of conformal blocks. The space of conformal blocks forms a vector bundle over the configuration space of points on the Riemann sphere. It will be shown that this vector bundle admits a natural flat connection called the KZ connection.

Jones-Witten theory

The Jones polynomial is a polynomial invariant for knots and links discovered in the middle of the 1980's by V. Jones from the point of view of the theory of operator algebras (see [**28**]). In this book we define the Jones polynomial by means of the holonomy of the KZ connection. In a seminal article [**55**], E. Witten proposed new invariants of 3-manifolds based on Chern-Simons gauge theory and gave a new interpretation of the Jones polynomial.

Let M be an oriented closed 3-manifold and G a simply connected Lie group. We will deal with the case $G = SU(2)$. We consider a principal G bundle over M. Since G is simply connected, P is a topologically trivial G bundle over M. We denote by $\mathcal{A}_M$ the space of connections on P. We see that $\mathcal{A}_M$ is an affine space. Taking a trivial connection as a base point, we identify $\mathcal{A}_M$ with $\Omega^1(M, \mathfrak{g})$, the space of 1-forms on M with values in $\mathfrak{g}$. Here $\mathfrak{g}$ is the Lie algebra of G. The gauge group $\mathcal{G}$ of P is identified with the space of smooth maps from M to G and acts on $\mathcal{A}_M$ by the pull-back. For $A \in \mathcal{A}_M$

we set

$$CS(A) = \frac{1}{8\pi^2} \int_M \mathrm{Tr}\left(A \wedge dA + \frac{2}{3} A \wedge A \wedge A\right), \ A \in \mathcal{A}_M,$$

and call CS the *Chern-Simons functional*. The invariant proposed by Witten is formally written as

$$Z_k(M) = \int_{\mathcal{A}_M/\mathcal{G}} \exp\left(2\pi\sqrt{-1}kCS(A)\right) \mathcal{D}A$$

using Feynman's path integral. Here k is an integer. In order to reformulate Witten's invariant in a rigorous setting we develop a theory for a 3-manifold with boundary. Here the space of conformal blocks in conformal field theory will play a role as the quantum Hilbert space. For a 3-manifold with boundary Σ we will determine a vector in the quantum Hilbert space associated with Σ. The holonomy of the space of conformal blocks describes the gluing rule. In this way, given a description of a closed oriented manifold by a Dehn surgery or a Heegaard splitting, we can establish a definition of $Z_k(M)$.

Let us sketch the above construction from a geometric point of view. For an oriented 3-manifold M with boundary Σ we take a G connection on Σ and consider the subspace $\mathcal{A}_{M,\alpha}$ of $\mathcal{A}_M$ of any connection whose restriction on Σ coincides with α. We denote by $Z_k(M)_\alpha$ the above formal expression of Feynman's path integral integrated over $\mathcal{A}_{M,\alpha}$. This formal expression of $Z_k(M)_\alpha$ should live in the space of G connections on Σ modulo the action of the gauge group. Computing the action of the gauge group on the Chern-Simons functional, we see that $Z_k(M)_\alpha$ should be a section of a complex line bundle over the moduli space of flat G bundles over Σ. Furthermore, by means of the Kähler polarization, we obtain the space of holomorphic sections of this complex line bundle as a quantum Hilbert space. It turns out that this quantum Hilbert space coincides with the space of conformal blocks. An important point is that the space of conformal blocks is a finite dimensional complex vector space. Our starting point was an infinite dimensional object, the space of connections over a 3-manifold. The above procedure permits us to extract a finite dimensional object out of this infinite dimensional one. Let us decompose a closed 3-manifold into two parts with common boundary Σ. The invariant $Z_k(M)$ is expressed by looking at the action of the gluing map on the space of conformal blocks associated with Σ. This leads us to an expression of $Z_k(M)$ based on a Dehn surgery or a

Heegaard splitting of M mentioned above. A main theme of this book is to extract such finite invariants from infinite dimensional objects.

Chern-Simons perturbation theory

In Chapter 3, we focus on another approach to the above infinite dimensional integral called the Chern-Simons perturbation theory. Let us explain an analogy in the finite dimensional case. We are going to consider the asymptotic behaviour of the integral

$$Z_k = \int_{\mathbf{R}^n} e^{\sqrt{-1}kf(x_1,\cdots,x_n)} \, dx_1 \cdots dx_n$$

as $k \to \infty$. Under a certain assumption the asymptotic behaviour of the above integral is expressed as the sum of contributions coming from critical points of f. The principal term of the asymptotic expansion is given by the Hessian of a critical point. The higher order terms are described by Feynman diagrams. We apply such an investigation to the case of the partition function of the Chern-Simons functional and construct topological invariants of 3-manifolds expressed by integrals of Green forms.

As the coefficients of the perturbative expansion of the Jones polynomial, we obtain a series of finite type invariants for knots. These invariants will be treated systematically in the framework of Vassiliev invariants. We give a universal integral expression for such invariants using the Kontsevich integral.

Notes

For more details about basic physical concepts described here, we refer the reader to [47]. An exposition on the geometric quantization can be found in Woodhouse [57]. In this book we mainly focus on topological invariants for knots and 3-manifolds arising from conformal field theory based on Witten's idea in the article [55]. The reader will find an introductory treatment on this subject in Atiyah [5]. We will stress geometric aspects of the theory and will not mention a method based on quantum groups. See, for example, Kassel [31] for a relation to the theory of quantum groups. In Chapter 1, we will deal with loop groups. A basic reference on loop groups is [46]. For a formulation of conformal field theory based on loop groups we mainly follow Gawedzki [24] (see also [37]). We refer the reader to Kac [29] for the representation theory of affine Lie algebras.

CHAPTER 1

Geometric Aspects
of Conformal Field Theory

Conformal field theory was initiated by Belavin, Polyakov and Zamolodchikov in the pioneering paper [**13**]. It is a quantum field theory describing infinite dimensional symmetries of critical phenomena in two dimensions. In this chapter we focus on geometric aspects of conformal field theory. We start from the Wess-Zumino-Witten model and describe how it gives rise to conformally invariant quantum field theory. In this construction an infinite dimensional Lie group called a loop group and a complex line bundle over the loop group will play a central role.

First, in Section 1.1, we will briefly review central extensions of loop groups and their infinitesimal versions called loop algebras. We will construct complex line bundles over the loop groups. In Section 1.2, we recall basic facts about affine Lie algebras and their representations. In particular, we deal with integrable highest weight modules over affine Lie algebras and the action of the Virasoro Lie algebra using Sugawara forms. In Section 1.3, we formulate the Wess-Zumino-Witten model. For a smooth map f from a Riemann surface Σ to a Lie group G we define an action $S(f)$. The main ingredient of the Wess-Zumino-Witten model is to compute the partition function of the above action, that is, the "average" of $\exp(-S(f))$ for all smooth maps $f : \Sigma \to G$. Such a partition function was originally formulated in terms of Feynman's path integral. In this book we develop a geometric approach. We consider a Riemann surface with boundary and show that in this case $\exp(-S(f))$ is regarded as an element of a fibre of a complex line bundle over the loop group. The partition function should be a section of this complex line bundle with a certain conformal invariance. Since the space of sections of the above complex line bundle admits an action of the affine Lie algebra, we can formulate its subspace in which the partition function lies, based on representations of affine Lie algebra. This subspace will be called the space of conformal blocks and we give its algebraic definition in Section 1.4. The space of conformal blocks forms a vector bundle over the moduli space of Riemann surfaces with a projectively flat connection. In this book we will mainly deal with the case of the Riemann sphere. In Section 1.5 we will show that in the case of the Riemann sphere the space of conformal blocks forms a vector bundle over the configuration space of distinct points on the Riemann sphere. Furthermore, we will describe a natural flat connection on the above vector bundle called the Knizhnik-Zamolodchikov connection. Section 1.6 will provide a dictionary describing a relation between the above geometric approach and physicists' approach using vertex operators and operator product expansions. Here we will mention briefly the space of conformal blocks of a torus and the Verlinde formula for the dimension of the space of conformal blocks.

1.1. Loop groups and affine Lie algebras

Let G be a compact connected Lie group. We denote by LG the space of smooth maps from $S^1 = \{z \in \mathbf{C} \mid |z| = 1\}$ to G. We introduce a group structure on LG by defining

$$(\gamma_1 \cdot \gamma_2)(z) = \gamma_1(z)\gamma_2(z), \quad \gamma_1, \gamma_2 \in LG.$$

The product $(\gamma_1, \gamma_2) \mapsto \gamma_1 \cdot \gamma_2$ is a smooth map as is the operation of inversion $\gamma \mapsto \gamma^{-1}$. This means that LG is an infinite dimensional Lie group. We call LG the *loop group* of G. In this chapter we will deal with the case when the Lie group G is $SU(2)$.

The loop algebra is the complexified Lie algebra of LG. In this book we adopt a formal version using Laurent series. Let $\mathbf{C}((t))$ denote the $\mathbf{C}$ algebra of the Laurent series expressed as

$$f(t) = \sum_{n=-m}^{\infty} a_n t^n$$

with some integer m. Let $\mathfrak{g}$ be the complexified Lie algebra of the Lie group G. In the case G is $SU(2)$, $\mathfrak{g}$ is the Lie algebra $sl_2(\mathbf{C})$. We set

$$L\mathfrak{g} = \mathfrak{g} \otimes \mathbf{C}((t)).$$

With the Lie bracket defined by

(1.1) $$[X \otimes f, Y \otimes g] = [X, Y] \otimes fg$$

for $X \otimes f, Y \otimes g \in L\mathfrak{g}$, we see that $L\mathfrak{g}$ is a complex Lie algebra.

Now we describe a central extension of $L\mathfrak{g}$. As a vector space we define $\widehat{\mathfrak{g}}$ to be the direct sum

$$L\mathfrak{g} \oplus \mathbf{C}c$$

of $L\mathfrak{g}$ and the one dimensional complex vector space with basis c. We will equip $\widehat{\mathfrak{g}}$ with the structure of a Lie algebra such that c belongs to the center of $L\mathfrak{g}$, i.e., $[c, \xi] = 0$ for any $\xi \in \widehat{\mathfrak{g}}$. Using a bilinear form $\omega : L\mathfrak{g} \times L\mathfrak{g} \to \mathbf{C}$, we define a bracket for $\widehat{\mathfrak{g}}$ by

$$[\xi + \alpha c, \eta + \beta c] = [\xi, \eta] + \omega(\xi, \eta)c, \quad \xi, \eta \in L\mathfrak{g}, \quad \alpha, \beta \in \mathbf{C}.$$

The above bracket defines a structure of a Lie algebra on $\widehat{\mathfrak{g}}$ if and only if the following conditions (1.2) and (1.3) are satisfied for any $x, y, z \in L\mathfrak{g}$.

(1.2) $$\omega(x, y) = -\omega(y, x).$$
(1.3) $$\omega([x, y], z) + \omega([y, z], x) + \omega([z, x], y) = 0.$$

In fact the condition (1.2) is equivalent to the anti-symmetry $[\xi, \eta] = -[\eta, \xi]$ and the condition (1.3) is equivalent to the Jacobi identity. With such a bilinear form ω we define the Lie bracket for $\widehat{\mathfrak{g}}$ by

$$[X \otimes f, Y \otimes g] = [X, Y] \otimes fg + \omega(X \otimes f, Y \otimes g)c.$$

The condition (1.3) for ω is called a *2-cocycle condition*. Two bilinear forms ω and ω' satisfying the conditions (1.2) and (1.3) define isomorphic Lie algebra structures on $L\mathfrak{g} \oplus \mathbf{C}c$ if and only if there exists a linear map $\mu : L\mathfrak{g} \to \mathbf{C}$ such that

$$\omega(x, y) = \omega'(x, y) + \mu([x, y])$$

for any $x, y \in L\mathfrak{g}$. In this way, associated with a bilinear form ω satisfying the conditions (1.2) and (1.3), we can equip $\widehat{\mathfrak{g}} = L\mathfrak{g} \oplus \mathbf{C}c$ with a structure of a Lie algebra so that c belongs to the center. This Lie algebra $\widehat{\mathfrak{g}}$ is called a *central extension* of $L\mathfrak{g}$.

Let us recall the notion of the cohomology of Lie algebras. For a Lie algebra $\mathfrak{a}$ and a left $\mathfrak{a}$ module M the p-th cochain group of $\mathfrak{a}$ with coefficients in M is by definition

$$C^p(\mathfrak{a}, M) = \mathrm{Hom}_{\mathbf{C}}(\bigwedge^p \mathfrak{a}, M).$$

The differential $d_p : C^p(\mathfrak{a}, M) \to C^{p+1}(\mathfrak{a}, M)$ is defined by

$$(d\omega)(x_0, x_1, \cdots, x_p)$$

$$= \sum_{i=0}^{p} (-1)^i \, x_i \, \omega(x_0, \cdots, \widehat{x_i}, \cdots, x_p)$$

$$+ \sum_{0 \leq i < j \leq p} (-1)^{i+j} \omega([x_i, x_j], x_0, \cdots, \widehat{x_i}, \cdots, \widehat{x_j}, \cdots, x_p)$$

for $\omega \in C^p(\mathfrak{a}, M)$. Here the notation $\widehat{x_i}$ means that we delete the i-th component x_i. The cohomology of the above complex

$$H^p(\mathfrak{a}, M) = \mathrm{Ker} \, d_p / \mathrm{Im} \, d_{p-1}$$

is called the p-th cohomology of the Lie algebra $\mathfrak{a}$ with coefficients in M.

In terms of the Lie algebra cohomology the above construction of a central extension of $L\mathfrak{g}$ based on a bilinear form ω shows that for any element of $H^2(L\mathfrak{g}, \mathbf{C})$ we can associate an isomorphism class of a central extension of the loop algebra $L\mathfrak{g}$. Here we regard $\mathbf{C}$ as a trivial $\mathfrak{g}$ module. Conversely, suppose that

$$0 \to \mathbf{C} \to \widetilde{L\mathfrak{g}} \xrightarrow{\pi} L\mathfrak{g} \to 0$$

is a central extension of $L\mathfrak{g}$. Choose a section $s : L\mathfrak{g} \to \widetilde{L\mathfrak{g}}$ as vector spaces. Namely, s is a linear map satisfying $\pi \circ s = id$. For $Z \in L\mathfrak{g}$ we write $s(Z) = \widetilde{Z}$ and set

$$\eta(X, Y) = [\widetilde{X}, \widetilde{Y}] - \widetilde{[X, Y]}, \quad X, Y \in L\mathfrak{g}.$$

It can be shown that η determines a cohomology class in $H^2(L\mathfrak{g}, \mathbf{C})$ which does not depend on the choice of a section s. Thus we have shown that there is a one-to-one correspondence between the set of isomorphism classes of the central extensions of $L\mathfrak{g}$ and $H^2(L\mathfrak{g}, \mathbf{C})$. This is a well-known fact for general Lie algebras. We denote by

$$\langle \, , \, \rangle : \mathfrak{g} \times \mathfrak{g} \to \mathbf{C}$$

a *Cartan-Killing form* of the Lie algebra $\mathfrak{g}$. By definition it is a non-degenerate symmetric bilinear form invariant under the adjoint action of $\mathfrak{g}$, i.e.,

$$\langle [X, Y], Z \rangle = \langle X, [Y, Z] \rangle$$

for any $X, Y, Z \in \mathfrak{g}$. In our case of $\mathfrak{g} = sl_2(\mathbf{C})$, we fix a Cartan-Killing form defined by

$$\langle X, Y \rangle = \mathrm{Tr}(XY).$$

By means of the Cartan-Killing form we can construct a 2-cocycle of the loop algebra $L\mathfrak{g}$ as follows.

PROPOSITION 1.1. *For the loop algebra $L\mathfrak{g}$ we have $H^2(L\mathfrak{g}, \mathbf{C}) \cong \mathbf{C}$. The cohomology $H^2(L\mathfrak{g}, \mathbf{C})$ has a basis represented by the 2-cocycle ω defined by*

$$\omega(X \otimes f, Y \otimes g) = \langle X, Y \rangle \operatorname{Res}_{t=0}(dfg)$$

where $\operatorname{Res}_{t=0}(\sum_n c_n t^n dt) = c_{-1}$.

PROOF. First we show that any cohomology class in $H^2(L\mathfrak{g}, \mathbf{C})$ is represented by a 2-cocycle invariant under conjugation by the Lie group G. We regard G as the set of constant loops in LG and we express $g \in G$ as $\exp tZ, Z \in \mathfrak{g}$. For $X \in L\mathfrak{g}$ we have $gXg^{-1} = X + t[Z, X] + O(t^2)$. Let α be a 2-cocycle of $L\mathfrak{g}$. We have

$$\lim_{t \to 0} \frac{1}{t} \left[\alpha(gXg^{-1}, gYg^{-1}) - \alpha(X, Y) \right] = \alpha(Z, [X, Y])$$

for $X, Y \in L\mathfrak{g}$ by means of the 2-cocycle condition for α. By defining a 1-cochain $\mu_Z : L\mathfrak{g} \to \mathbf{C}$ by $\mu_Z(U) = \alpha(Z, U)$ for $Z \in L\mathfrak{g}$ we can express the right hand side of the above equation as $\mu_Z([X, Y])$. Therefore the cohomology class of α is invariant under the infinitesimal conjugation. For $g \in G$ we denote by α_g the 2-cocycle of $L\mathfrak{g}$ defined by $\alpha_g(X, Y) = \alpha(gXg^{-1}, gYg^{-1})$. The 2-cocycle

$$\int_G \alpha_g \, dg$$

obtained by averaging α_g over the compact Lie group G is invariant under conjugation by G and is cohomologous to α by the above argument. Thus we may suppose that the 2-cocycle α is invariant under conjugation by G.

Let α be a 2-cocycle $\alpha : \bigwedge^2 L\mathfrak{g} \to \mathbf{C}$ invariant under conjugation by G. We set $\alpha_{m,n}(X, Y) = \alpha(X \otimes t^m, Y \otimes t^n)$ for $X, Y \in \mathfrak{g}$. We see that $\alpha_{m,n} : \mathfrak{g} \times \mathfrak{g} \to \mathbf{C}$ is a bilinear form and is invariant under the adjoint action of $\mathfrak{g}$. The bilinear form $\alpha_{m,n}$ is necessarily symmetric since the Lie algebra $\mathfrak{g}$ is simple. Combining with the anti-symmetry of α we have $\alpha_{m,n} = -\alpha_{n,m}$. The 2-cocycle condition for α is expressed as

$$\alpha_{m+n,p} + \alpha_{n+p,m} + \alpha_{p+m,n} = 0.$$

Putting $n = p = 0$ we find $\alpha_{m,0} = 0$ for all m. Putting $p = -m - n$ we have $\alpha_{m+n,-m-n} = \alpha_{m,-m} + \alpha_{n,-n}$, whence $\alpha_{m,-m} = m\alpha_{1,-1}$. Putting $p = q - m - n$ we have $\alpha_{q-m-n,m+n} = \alpha_{q-m,m} + \alpha_{q-n,n}$, whence $\alpha_{q-k,k} = k\alpha_{q-1,1}$. This implies that $\alpha_{m,n} = 0$ if $m + n \neq 0$, for $q\alpha_{q-1,1} = \alpha_{0,q} = 0$. Thus we have

$$\alpha_{m,n} = m\delta_{m+n,0}\alpha_{1,-1}.$$

Since $\alpha_{1,-1} : \mathfrak{g} \times \mathfrak{g} \to \mathbf{C}$ is a $\mathfrak{g}$ invariant symmetric bilinear form it is equal to the Cartan-Killing form up to a constant multiple. Setting $\alpha_{1,-1}$ as the Cartan-Killing form we obtain the desired 2-cocycle ω. The fact that the 2-cocycle ω is not a coboundary is shown as follows. Take H in the Cartan subalgebra of $\mathfrak{g}$ such that $\langle H, H \rangle \neq 0$. Suppose ω is a coboundary. Then there exists a 1-cocycle $\beta : L\mathfrak{g} \to \mathbf{C}$ such that $\omega(H \otimes t, H \otimes t^{-1}) = \beta([H \otimes t, H \otimes t^{-1}])$. We have $\omega(H \otimes t, H \otimes t^{-1}) = \langle H, H \rangle \neq 0$. On the other hand, we have $[H \otimes t, H \otimes t^{-1}] = 0$ in $L\mathfrak{g}$. This is a contradiction, which implies that ω is not a coboundary. This completes the proof. $\qquad\qquad\square$

The central extension $\widehat{\mathfrak{g}}$ of $\mathfrak{g}$ defined by the 2-cocycle ω is called the *affine Lie algebra* associated with $\mathfrak{g}$. The Lie bracket of $\widehat{\mathfrak{g}}$ is expressed as

$$(1.4) \qquad [X \otimes t^m, Y \otimes t^n] = [X, Y] \otimes t^{m+n} + \langle X, Y \rangle m\delta_{m+n,0}c$$

for $X, Y \in \mathfrak{g}$.

We denote by $\mathrm{Diff}(S^1)$ the group of diffeomorphisms of the unit circle $S^1 = \{z \in \mathbf{C} \mid |z| = 1\}$. The group $\mathrm{Diff}(S^1)$ has a structure of an infinite dimensional Lie group. The Lie algebra of $\mathrm{Diff}(S^1)$ is identified with the Lie algebra of smooth vector fields on S^1 denoted by $\mathrm{Vect}\, S^1$. We will consider the vector fields on S^1 which can be extended over $\mathbf{C} \setminus \{0\}$ as

$$f(z)\frac{d}{dz}$$

with a Laurent polynomial $f(z)$. We set

$$A = \left\{ f(z)\frac{d}{dz} \;\middle|\; f(z) \in \mathbf{C}[z, z^{-1}] \right\}$$

where $\mathbf{C}[z, z^{-1}]$ stands for the $\mathbf{C}$ algebra of Laurent polynomials. The algebra A is equipped with a structure of a Lie algebra with the Lie bracket defined by

$$\left[f(z)\frac{d}{dz}, g(z)\frac{d}{dz} \right] = \{f(z)g'(z) - g(z)f'(z)\}\frac{d}{dz}.$$

This is an infinite dimensional Lie algebra with basis

$$L_n = -z^{n+1}\frac{d}{dz}, \quad n \in \mathbf{Z}.$$

The restriction of the vector field L_n on S^1 is written as

$$L_n = ie^{in\theta}\frac{\partial}{\partial z}$$

with $z = e^{i\theta}$. With respect to the above Lie bracket of A the vector fields $L_n, n \in \mathbf{Z}$, satisfy

(1.5) $$[L_m, L_n] = (m - n)L_{m+n}.$$

The vector field L_n generates an infinitesimal transformation of $\mathbf{C} \setminus \{0\}$ given by

$$\varphi_t(z) = z - tz^{n+1}.$$

Among $L_n, n \in \mathbf{Z}$, only L_0, L_{-1} and L_1 are extended smoothly on the Riemann sphere $\mathbf{C}P^1$. They satisfy the relations

$$[L_0, L_1] = -L_1, \quad [L_0, L_{-1}] = L_{-1}, \quad [L_1, L_{-1}] = 2L_0$$

and generate the Lie algebra $sl_2(\mathbf{C})$.

As in the case of affine Lie algebras we construct a central extension of A starting from a 2-cocycle $\alpha : A \times A \to \mathbf{C}$. We have the following proposition.

PROPOSITION 1.2. *For the Lie algebra of Laurent polynomial vector fields A we have $H^2(A, \mathbf{C}) \cong \mathbf{C}$. The cohomology $H^2(A, \mathbf{C})$ has a basis represented by the 2-cocycle α defined by*

$$\omega\left(f\frac{d}{dz}, g\frac{d}{dz}\right) = \frac{1}{12}\mathrm{Res}_{t=0}(f'''g\,dz).$$

PROOF. Let α be a 2-cocycle of A and put $\alpha_{p,q} = \alpha(L_p, L_q)$. The 2-cocycle condition for (L_0, L_p, L_q) shows that the cohomology class of α is not changed by rotation. By averaging over the circle we may assume that α is invariant by rotation. Then we have $\alpha_{p,q} = 0$ if $p + q \neq 0$. Putting $\alpha_p = \alpha_{p,-p}$ the 2-cocycle condition gives

$$(p + 2q)\alpha_p - (2p + q)\alpha_q = (p - q)\alpha_{p+q}$$

since $\alpha_{-p} = -\alpha_p$. Thus we see that α_p is expressed as $\alpha_p = \lambda p^3 + \mu p$. The cohomology class of α does not depend on the value of μ. We set $\lambda = \frac{1}{12}, \mu = -\frac{1}{12}$ and we obtain the desired 2-cocycle. $\square$

By means of the 2-cocycle α in Proposition 1.2 we define the Lie bracket on $\mathcal{V} = A \oplus \mathbf{C}c$ by

$$\left[f\frac{d}{dz} + \xi c, \ g\frac{d}{dz} + \eta c \right] = \left[f\frac{d}{dz}, g\frac{d}{dz} \right] + \alpha \left(f\frac{d}{dz}, g\frac{d}{dz} \right) c, \quad \xi, \eta \in \mathbf{C}.$$

The Lie algebra $\mathcal{V}$ is a central extension of A and is called the *Virasoro Lie algebra*. The Lie bracket of the Virasoro Lie algebra may be expressed as

$$(1.6) \qquad [L_m, L_n] = (m-n)L_{m+n} + \frac{m^3 - m}{12} \delta_{m+n,0} \ c.$$

We deal with the case $G = SU(2)$. The complexification $G_{\mathbf{C}}$ of G is $G_{\mathbf{C}} = SL(2, \mathbf{C})$. Our next object is to study complex line bundles over the loop group $LG_{\mathbf{C}}$ and central extensions of $LG_{\mathbf{C}}$. These will play an essential role in the Wess-Zumino-Witten model as we will see in Section 1.3.

We recall basic facts on complex line bundles. See [**57**] for a more detailed description. Let M be a smooth manifold and let L be a complex line bundle over M. We fix a connection ∇ on the complex line bundle L. Suppose that locally the complex line bundle L is trivialized over an open set U of M as $U \times \mathbf{C}$ and express the connection ∇ as $\nabla = d - 2\pi\sqrt{-1}\alpha_U$ over U, where α_U is a 1-form defined over U. Let $\{U_\lambda\}, \lambda \in \Lambda$, be an open covering of M so that the line bundle L is trivialized on each U_λ. For each U_λ we associate the 1-form α_{U_λ} defined in the above way. It turns out that $d\alpha_{U_\lambda}, \lambda \in \Lambda$, define globally a 2-form on M. This 2-form is called the *first Chern form* of the connection ∇ and is denoted by $c_1(\nabla)$. We have $dc_1(\nabla) = 0$ and that the cohomology class of $c_1(\nabla)$ does not depend on the choice of a connection ∇. Moreover, the cohomology class lies in the image of the natural map $i : H^2(M, \mathbf{Z}) \to H^2(M, \mathbf{R})$. This cohomology class is the *first Chern class* of L.

We now suppose that the manifold M is simply connected and fix a base point $x_0 \in M$. Let $\gamma : [0, 1] \to M$ be a smooth loop such that $\gamma(0) = \gamma(1) = x_0$. The pull back γ^*L is a complex line bundle over the unit interval $[0, 1]$ with the induced connection $\gamma^*\nabla$. A *horizontal section* s of γ^*L is a section satisfying $(\gamma^*\nabla)s = 0$. For u in the fibre of γ^*L over 0, we associate the horizontal section of γ^*L specified by $s(0) = u$. Since M is simply connected we have an oriented 2-dimensional disc D with boundary γ. By means of the

Stokes theorem we see that $s(1)$ is given by

$$u \exp \left(2\pi\sqrt{-1} \int_D c_1(\nabla) \right).$$

We denote by L_{x_0} the fibre of L over x_0. By associating $s(1)$ to $u = s(0)$ we obtain a linear transformation of L_{x_0}, which is called the *holonomy* of ∇ around the loop γ. Let us notice that the natural map $i : H^2(M, \mathbf{Z}) \rightarrow H^2(M, \mathbf{R})$ is injective because M is simply connected. We have the following proposition.

PROPOSITION 1.3. *Let M be a simply connected smooth manifold and ω a closed 2-form on M whose cohomology class lies in the image of the natural injective map*

$$i : H^2(M, \mathbf{Z}) \rightarrow H^2(M, \mathbf{R}).$$

Then, there exists a complex line bundle L over M and a connection ∇ on L such that the first Chern form $c_1(\nabla)$ coincides with ω.

PROOF. We construct a complex line bundle L in the following way. We denote by $P_{x_0}(M)$ the set of smooth paths $p : [0, 1] \rightarrow M$ with $p(0) = x_0$. We introduce an equivalence relation $\sim$ on the Cartesian product $P_{x_0}(M) \times \mathbf{C}$ by defining $(p, u) \sim (q, v)$ if and only if the conditions $p(1) = q(1)$ and

$$v = u \exp \left(2\pi\sqrt{-1} \int_D \omega \right)$$

are satisfied. Here D is an oriented 2-dimensional disc in M whose boundary is the loop p followed by q^{-1}. From the hypothesis that ω determines a cohomology class in $H^2(M, \mathbf{Z})$ it follows that the above equivalence relation is well defined, since it does not depend on the choice of a disc D. We set

$$L = P_{x_0}(M) \times \mathbf{C}/ \sim$$

and define the projection map $\pi : L \rightarrow M$ by $\pi(p, u) = p(1)$. We see that L has a structure of a complex line bundle over M. From the definition of the equivalence relation $\sim$ we obtain a linear transformation of L_{x_0} for each loop γ based at x_0 by associating $u \exp \left(2\pi\sqrt{-1} \int_D \omega \right)$ to $u \in L_{x_0}$ where D is an oriented 2-dimensional disc in M bounded by the loop γ. We have a connection ∇ for the line bundle L whose holonomy is given by the above linear transformation of L_{x_0}. Comparing with the description of the holonomy for the connection ∇, we see that the first Chern form of ∇ coincides with ω. □

Let us now proceed to describe complex line bundles over the loop group $LG_{\mathbf{C}}$. First we show that the loop group LG is simply connected. We denote by ΩG the subset of LG consisting of the loops $\gamma : S^1 \to G$ such that $\gamma(1)$ is the unit element of G. The Lie group G is embedded in LG as the set of constant loops. Since LG is homeomorphic to the Cartesian product $G \times \Omega G$, we have

$$\pi_1(LG) \cong \pi_1(G) \oplus \pi_1(\Omega G) \cong \pi_1(G) \oplus \pi_2(G).$$

The Lie group G is simply connected and we have $\pi_2(G) = 0$, which implies that LG is simply connected. Similarly, we see that the loop group $LG_{\mathbf{C}}$ is simply connected.

Let us suppose that ω is a closed 2-form over $LG_{\mathbf{C}}$ whose cohomology class lies in $H^2(LG_{\mathbf{C}}, \mathbf{Z})$. By applying a method in the proof of Proposition 1.3 we can construct a complex line bundle over $LG_{\mathbf{C}}$ associated with ω. An explicit expression of such a 2-form ω is given as follows. The Lie algebra $\mathrm{Lie}(LG_{\mathbf{C}})$ of the loop group $LG_{\mathbf{C}}$ is identified with the set of smooth maps from S^1 to $\mathfrak{g}$. Here $\mathfrak{g} = sl_2(\mathbf{C})$ is the Lie algebra of $G_{\mathbf{C}}$. Considering the Fourier expansion of a smooth map from S^1 to $\mathfrak{g}$, we define a 2-cocycle of $\mathrm{Lie}(LG_{\mathbf{C}})$ as in Proposition 1.1 by the formula

$$(1.7) \qquad \omega(\xi, \eta) = \frac{1}{4\pi^2} \int_0^{2\pi} \langle \xi'(\theta), \eta(\theta) \rangle \, d\theta$$

for $\xi, \eta \in \mathrm{Lie}(LG_{\mathbf{C}})$. One can check that ω defines a 2-cocycle of $\mathrm{Lie}(LG_{\mathbf{C}})$. This 2-cocycle ω determines a left invariant 2-form on the loop group $LG_{\mathbf{C}}$ and by means of the 2-cocycle condition one can show that it is a closed form. We denote this 2-form also by ω.

We are going to show that the de Rham cohomology class of the above 2-form ω lies in the integral cohomology group $H^2(LG_{\mathbf{C}}, \mathbf{Z})$. For $X \in SU(2)$ the 1-form $\mu = X^{-1}dX$ takes values in the Lie algebra $su(2)$ and is invariant under the action of the left multiplication by G. The 1-form μ is called the *Maurer-Cartan form* of the Lie group $SU(2)$. The de Rham cohomology class of the 3-form

$$(1.8) \qquad \sigma = \frac{1}{24\pi^2} \mathrm{Tr}(\mu \wedge \mu \wedge \mu)$$

generates the integral cohomology group $H^3(SU(2), \mathbf{Z})$. Let $\phi : LG \times S^1 \to G$ be the map defined by $\phi(\gamma, z) = \gamma(z)$ and put

$$\omega_0 = - \int_{S^1} \phi^* \sigma.$$

It follows from the definition that ω_0 satisfies $d\omega_0 = 0$ and that the de Rham cohomology class of ω_0 lies in the integral cohomology group $H^2(LG, \mathbf{Z})$. The 2-form ω_0 defined on LG extends naturally over $LG_{\mathbf{C}}$.

The tangent space of $LG_{\mathbf{C}}$ at $\gamma \in LG_{\mathbf{C}}$ is denoted by $\mathrm{T}_\gamma LG_{\mathbf{C}}$. It is identified with the set of left invariant vector fields on $LG_{\mathbf{C}}$ and is regarded as the Lie algebra $\mathrm{Lie}(LG_{\mathbf{C}})$. The 1-form β on $LG_{\mathbf{C}}$ is defined by the formula

$$\beta_\gamma(\xi) = \frac{1}{8\pi^2} \int_0^{2\pi} \langle \gamma(\theta)^{-1} \gamma'(\theta), \gamma(\theta)^{-1} \xi(\theta) \rangle \, d\theta$$

for $\gamma \in LG_{\mathbf{C}}$ and $\xi \in \mathrm{T}_\gamma LG_{\mathbf{C}}$. It can be shown that $\omega = \omega_0 + d\beta$ holds. Hence ω defines an integral cohomology class as well as ω_0. Summarizing the above argument, we can conclude that we have a complex line bundle with a connection over $LG_{\mathbf{C}}$ whose first Chern form is the 2-form ω. This complex line bundle is called the fundamental line bundle over $LG_{\mathbf{C}}$. It is known that we have an isomorphism $H^2(LG, \mathbf{Z}) \cong \mathbf{Z}$ and that this cohomology group is spanned by the de Rham cohomology class of ω defined by the formula (1.7).

1.2. Representations of affine Lie algebras

Let $\mathfrak{g}$ be a complex Lie algebra and V a complex vector space. The set of linear transformations of V is denoted by $\mathrm{End}(V)$. A linear representation of $\mathfrak{g}$ on V is a linear map $\rho : \mathfrak{g} \to \mathrm{End}(V)$ such that

$$\rho([X, Y]) = \rho(X)\rho(Y) - \rho(Y)\rho(X)$$

for any $X, Y \in \mathfrak{g}$. The vector space V is called a left $\mathfrak{g}$ module or simply a $\mathfrak{g}$ module. For $X \in \mathfrak{g}$ and $v \in V$, $\rho(X)v$ will also be denoted by Xv. We recall basic facts on representations of affine Lie algebras which will be needed in this book. We refer the reader to Kac [29] for details. We deal with the case $\mathfrak{g} = sl_2(\mathbf{C})$. The central extension $\widehat{\mathfrak{g}}$ of the loop algebra $L\mathfrak{g}$ is called the affine Lie algebra of type $A_1^{(1)}$.

First we quickly review representations of $\mathfrak{g} = sl_2(\mathbf{C})$. Let us recall that the Lie algebra $sl_2(\mathbf{C})$ consists of 2 by 2 complex matrices with trace 0. The Lie bracket is defined by

$$[X, Y] = XY - YX, \quad X, Y \in sl_2(\mathbf{C}).$$

As a complex vector space $sl_2(\mathbf{C})$ has as a basis

$$H = \begin{pmatrix} 1 & 0 \\ 0 & -1 \end{pmatrix}, \quad E = \begin{pmatrix} 0 & 1 \\ 0 & 0 \end{pmatrix}, \quad F = \begin{pmatrix} 0 & 0 \\ 1 & 0 \end{pmatrix}$$

and they satisfy the relations

$$[H, E] = 2E, \quad [H, F] = -2F, \quad [E, F] = H.$$

The Cartan-Killing form of $\mathfrak{g}$ is expressed as $\langle X, Y \rangle = \mathrm{Tr}(XY)$ for $X, Y \in \mathfrak{g}$. It is a non-degenerate symmetric invariant bilinear form. For the above basis H, E, F we have $\langle H, H \rangle = 2, \langle E, F \rangle = 1, \langle H, E \rangle = 0, \langle H, F \rangle = 0$. Let $\{I_\mu\}$ be an orthonormal basis of $\mathfrak{g}$ with respect to the Cartan-Killing form. We denote by $U(\mathfrak{g})$ the universal enveloping algebra of $\mathfrak{g}$. The Casimir element C in $U(\mathfrak{g})$ is defined by $C = \sum_\mu I_\mu I_\mu$. It is also expressed as

$$C = \frac{1}{2}H^2 + EF + FE.$$

We can verify that $[C, X] = 0$ holds for any $X \in U(\mathfrak{g})$.

Let λ be a complex number. A left $\mathfrak{g}$ module V is called a *highest weight representation* with highest weight λ if the following two conditions are satisfied.

1. There exists a non-zero vector $v \in V$ such that $Hv = \lambda v$ and $Ev = 0$ hold.
2. As a complex vector space V is generated by $F^n v, n = 0, 1, \cdots$.

The vector v as above is called a highest weight vector. If $F^n v$, $n = 0, 1, \cdots$, are linearly independent, then the above vector space V is called the *Verma module* with highest weight λ and is denoted by M_λ. Here, in the case of $sl_2(\mathbf{C})$, λ is set to be a complex number, but for a general Lie algebra, a highest weight is formulated as a linear form on its Cartan subalgebra.

For a generic complex number λ, the Verma module M_λ is shown to be an irreducible $\mathfrak{g}$ module, but for a special value λ, it can happen that M_λ is not irreducible. Consider the case when λ is a non-negative integer and set $\lambda = 2j$. We denote by V_λ the complex vector space of dimension $\lambda + 1$ with basis $u_m, m = j, j - 1, \cdots, -j + 1, -j$. The vector space V_λ is considered as a left $\mathfrak{g}$ module by

$$Hu_m = 2mu_m,$$

$$Eu_m = \sqrt{(j + m + 1)(j - m)}\, u_{m+1},$$

$$Fu_m = \sqrt{(j + m)(j - m + 1)}\, u_{m-1}.$$

Here u_j is a highest weight vector and we have $F^{\lambda+1}u_j = 0$. One can show that it is an irreducible representation of $\mathfrak{g}$. As a left $\mathfrak{g}$ module V_λ is isomorphic to the quotient of the Verma module M_λ by the submodule generated by $F^{\lambda+1}v$. We call V_λ the *spin j representation* of $\mathfrak{g}$. Since the Casimir element C lies in the center of the universal enveloping algebra $U(\mathfrak{g})$ it acts as a scalar on the irreducible representation V_λ. By computing the eigenvalue of C on the highest weight vector u_j, we see that this scalar is $2j(j+1)$. It is known that any finite dimensional representation of $sl_2(\mathbf{C})$ is equivalent to one of the above representations $V_\lambda, \lambda = 0, 1, \cdots$.

The right action of $\mathfrak{g}$ on the dual space $V_\lambda^* = \mathrm{Hom}_{\mathbf{C}}(V_\lambda, \mathbf{C})$ is specified uniquely by $\langle \xi X, \eta \rangle = \langle \xi, X\eta \rangle$ for any $\xi \in V_\lambda^*, \eta \in V_\lambda$ and $X \in \mathfrak{g}$. Here $\langle\,,\,\rangle$ stands for the natural pairing between V_λ^* and V_λ. The right $\mathfrak{g}$ module obtained in this way is irreducible and is called the right $\mathfrak{g}$ module with highest weight λ. We denote by $\{u_m^*\}$ the basis of V_λ^* dual to the basis $\{u_m\}$ of V_λ. With respect to the above right action we have $u_m^* H = 2m u_m^*, u_j^* F = 0$, and V_λ^* is spanned by the vectors $u_j^* E^n, n = 0, 1, \cdots, \lambda$.

By using the above right action, we define a left $\mathfrak{g}$ action ρ^* on V_λ^* by

$$\rho^*(X)\xi = -\xi X, \ X \in \mathfrak{g}, \xi \in V_\lambda^*.$$

We have

$$\rho^*([X,Y]) = \rho^*(X)\rho^*(Y) - \rho^*(Y)\rho^*(X)$$

and this shows that V_λ^* is also regarded as a left $\mathfrak{g}$ module by this action. We call it the *dual representation* of the representation of $\mathfrak{g}$ on V_λ. In the case $\mathfrak{g} = sl_2(\mathbf{C})$ V_λ is equivalent to its dual representation V_λ^*.

For left $\mathfrak{g}$ modules V and W, the tensor product $V \otimes W$ is regarded as a left $\mathfrak{g}$ module by

$$X(v \otimes w) = Xv \otimes w + v \otimes Xw, \quad X \in \mathfrak{g}, v \in V, w \in W.$$

This is called the tensor product of the representations V and W. In the case $\mathfrak{g} = sl_2(\mathbf{C})$, the way of decomposing the tensor product of two finite dimensional irreducible representations into irreducible ones is known as the *Clebsch-Gordan rule*.

PROPOSITION 1.4 (Clebsch-Gordan rule). *For finite dimensional irreducible representations* $V_{\lambda_1}, V_{\lambda_2}$ *and* V_{λ_3} *of* $sl_2(\mathbf{C})$, *we have*

$$\mathrm{Hom}_{sl_2(\mathbf{C})}(V_{\lambda_1} \otimes V_{\lambda_2} \otimes V_{\lambda_3}, \mathbf{C}) \cong \mathbf{C}$$

if and only if the condition

$$\lambda_1 + \lambda_2 + \lambda_3 \in 2\mathbf{Z},$$
$$|\lambda_1 - \lambda_2| \leq \lambda_3 \leq \lambda_1 + \lambda_2$$

is satisfied. Otherwise, we have

$$\mathrm{Hom}_{sl_2(\mathbf{C})}(V_{\lambda_1} \otimes V_{\lambda_2} \otimes V_{\lambda_3}, \mathbf{C}) = 0.$$

In particular, tensor products of $V = V_1$ are decomposed into irreducible representations as

$$V \otimes V = V_0 \oplus V_2,$$
$$V \otimes V \otimes V = 2V_1 \oplus V_3,$$
$$V \otimes V \otimes V \otimes V = 2V_0 \oplus 3V_2 \oplus V_4,$$
$$\cdots$$

Here the number inscribed before a vector space stands for the multiplicity of the representation. A general formula is given by means of binomial coefficients.

We briefly recall representations of affine Lie algebras. The subalgebra of $\mathbf{C}((t))$ consisting of the Laurent series of the form $\sum_{n>0} a_n t^n$ is denoted by A_+. Similarly, A_- stands for the subalgebra of $\mathbf{C}((t))$ consisting of the Laurent series of the form $\sum_{n<0} a_n t^n$. The Lie subalgebras N_+, N_0, N_- of the affine Lie algebra $\widehat{\mathfrak{g}}$ are defined respectively by

$$N_+ = [\mathfrak{g} \otimes A_+] \oplus \mathbf{C}E,$$
$$N_0 = \mathbf{C}H \oplus \mathbf{C}c,$$
$$N_- = [\mathfrak{g} \otimes A_-] \oplus \mathbf{C}F.$$

We have a direct sum decomposition

$$\widehat{\mathfrak{g}} = N_+ \oplus N_0 \oplus N_-$$

as Lie algebras.

Let k and λ be complex numbers. A left $\widehat{\mathfrak{g}}$ module $\widehat{V}_{k,\lambda}$ is called a *highest weight representation* with level k and highest weight λ if the following two conditions are satisfied.

1. There exists a non-zero vector v in $\widehat{V}_{k,\lambda}$ such that $N_+ v = 0$, $cv = kv$ and $Hv = \lambda v$ hold.
2. The $U(N_-)$ submodule of $\widehat{V}_{k,\lambda}$ generated by the vector v coincides with $\widehat{V}_{k,\lambda}$ where $U(N_-)$ denotes the universal enveloping algebra of N_-.

The first condition and the fact that c lies in the center of $\widehat{\mathfrak{g}}$ imply $cu = ku$ for any $u \in \widehat{V}_{k,\lambda}$. The vector v in the above conditions is called a highest weight vector of $\widehat{V}_{k,\lambda}$. By using the commutation relation (1.4) it can be shown that as a vector space $\widehat{V}_{k,\lambda}$ is generated by $a_1 \cdots a_r v$ with $a_1, \cdots, a_r \in N_-, r \geq 0$.

The rank 1 free $\widehat{\mathfrak{g}}$ module with level k and highest weight λ generated by the highest weight vector v is called the *Verma module* and is denoted by $M_{k,\lambda}$. Any representation $\widehat{V}_{k,\lambda}$ of $\widehat{\mathfrak{g}}$ with level k and highest weight λ is expressed as a quotient module of $M_{k,\lambda}$. That is, there exists a submodule J of $M_{k,\lambda}$ such that $\widehat{V}_{k,\lambda} = M_{k,\lambda}/J$. In particular, if J is a maximal proper submodule of $M_{k,\lambda}$, then we obtain as the quotient $M_{k,\lambda}/J$ an irreducible $\widehat{\mathfrak{g}}$ module. For general parameters k and λ the Verma module $M_{k,\lambda}$ is itself an irreducible $\widehat{\mathfrak{g}}$ module, but for special parameters it can happen that $M_{k,\lambda}$ contains a proper $\widehat{\mathfrak{g}}$ submodule. In the case when k is a positive integer the following fact is known by Kac.

PROPOSITION 1.5. *Let k be a positive integer and λ an integer such that $0 \leq \lambda \leq k$. Define $\chi \in M_{k,\lambda}$ by $\chi = (E \otimes t^{-1})^{k-\lambda+1} v$ where v is a highest weight vector of $M_{k,\lambda}$. Then, we have $N_+ \chi = 0$ and $U(N_-)\chi$ is a maximal proper submodule of $M_{k,\lambda}$. The quotient*

$$H_{k,\lambda} = M_{k,\lambda}/U(N_-)\chi$$

is an irreducible $\widehat{\mathfrak{g}}$ module.

Thus, for a positive integer k and an integer λ with $0 \leq \lambda \leq k$ we have constructed an irreducible $\widehat{\mathfrak{g}}$ module $H_{k,\lambda}$. This is called the *integrable highest weight module* with level k and highest weight λ. We refer the reader to [**29**] for the proof of Proposition 1.5. A highest weight λ satisfying the condition $0 \leq \lambda \leq k$ is called a highest weight of level k. The vector χ in Proposition 1.5 is called a *null vector*. The integrable highest weight module $H_{k,\lambda}$ contains as a subspace V_λ, a finite dimensional irreducible representation of $sl_2(\mathbf{C})$. When the level k is fixed, we simply write H_λ for $H_{k,\lambda}$.

The dual vector space H_λ^* of H_λ has the structure of a right $\widehat{\mathfrak{g}}$ module uniquely specified by $\langle \xi \alpha, \eta \rangle = \langle \xi, \alpha \eta \rangle$ for any $\xi \in H_\lambda^*, \eta \in H_\lambda$ and $\alpha \in \widehat{\mathfrak{g}}$. The dual vector $v^* \in H_\lambda^*$ of the highest weight vector $v \in H_\lambda$ satisfies $v^* N_- = 0$ with respect to the above right action and H_λ^* is generated by v^* as a right $U(N_+)$ module. By means of this right action of $\widehat{\mathfrak{g}}$ on H_λ^*, we define the left representation ρ^* of $\widehat{\mathfrak{g}}$ on H_λ^* by

$$\rho^*(X \otimes t^n)\xi = -\xi X \otimes t^{-n}, \ X \in \mathfrak{g}, \xi \in H_\lambda^*.$$

Thus, H_λ^* has a structure of a left $\widehat{\mathfrak{g}}$ module as well. This is called the *dual representation* of H_λ.

Our next object is an action of the Virasoro Lie algebra on H_λ. As we will see in Section 1.3 the left $\widehat{\mathfrak{g}}$ module H_λ is realized as a subspace of the space of sections of a certain complex line bundle over the loop group $LG_{\mathbf{C}}$. Since the group $\mathrm{Diff}(S^1)$ acts by the change of parameters on the space of smooth maps $f : S^1 \to G_{\mathbf{C}}$, it would be natural to expect that the Virasoro Lie algebra acts on H_λ. In fact the following *Sugawara operators* permit us to define an action of the Virasoro Lie algebra on H_λ. The Sugawara operator L_n in the case $n \neq 0$ is defined formally by

$$(1.9) \qquad L_n = \frac{1}{2(k+2)} \sum_{j \in \mathbf{Z}} \sum_\mu I_\mu \otimes t^{-j} \cdot I_\mu \otimes t^{n+j}$$

where $\{I_\mu\}$ is an orthonormal basis of $\mathfrak{g}$ with respect to the Cartan-Killing form. In the case $n = 0$, L_0 is defined by the formal sum

$$(1.10) \quad L_0 = \frac{1}{k+2} \sum_{j \geq 1} \sum_\mu I_\mu \otimes t^{-j} \cdot I_\mu \otimes t^j + \frac{1}{2(k+2)} \sum_\mu I_\mu \cdot I_\mu.$$

Notice that in the case $n \neq 0$, $I_\mu \otimes t^{-j}$ and $I_\mu \otimes t^{n+j}$ commute with each other. The above definition of L_n formally involves an infinite sum, but for each element ξ of $H_{k,\lambda}$, we see that $L_n \xi$ is expressed as a finite sum and L_n is a well-defined linear operator on $H_{k,\lambda}$. The Sugawara operator may be considered as an infinite dimensional extension of the Casimir element. By computing the Lie bracket $[L_m, I_\mu \otimes t^n]$ for the basis $\{I_\mu\}$ of $\mathfrak{g}$, we obtain the following proposition.

PROPOSITION 1.6. *The Sugawara operators L_m, $m \in \mathbf{Z}$, acting on the integrable highest weight $\widehat{\mathfrak{g}}$ module $H_{k,\lambda}$ satisfy the relation*

$$[L_m, X \otimes t^n] = -nX \otimes t^{m+n}, \quad X \in \mathfrak{g},$$

for any integer n.

Furthermore, performing a calculation based on Proposition 1.6, we have the following commutation relation.

PROPOSITION 1.7. *As linear operators on $H_{k,\lambda}$, the Sugawara operators L_m, $m \in \mathbf{Z}$, satisfy the commutation relation*

$$[L_m, L_n] = (m-n)L_{m+n} + \frac{m^3 - m}{12}\delta_{m+n,0}\frac{3k}{k+2}.$$

Thus, the integrable highest weight module $H_{k,\lambda}$ is regarded as a module over the Virasoro Lie algebra via the Sugawara operators. The action of the central element c of the Virasoro Lie algebra is the scalar multiplication by $\frac{3k}{k+2}$, which will be called the *central charge*.

With respect to the action of L_0, the integrable highest weight module $H_\lambda = H_{k,\lambda}$ is decomposed into its eigenspaces in the following manner. First, for $u \in V_\lambda$ we have

$$L_0 u = \frac{1}{2(k+2)} \left(\sum_\mu I_\mu \cdot I_\mu \right) u = \frac{j(j+1)}{k+2} u.$$

Here we set $\lambda = 2j$. We denote by Δ_λ the above eigenvalue $\frac{j(j+1)}{k+2}$ and call it the *conformal weight*. It follows from Proposition 1.6 that any eigenvalue of L_0 is expressed as $\Delta_\lambda + d$ with a non-negative integer d. We denote by $H_\lambda(d)$ the eigenspace with eigenvalue $\Delta_\lambda + d$. We have a direct sum decomposition

$$(1.11) \qquad\qquad H_\lambda = \bigoplus_{d \geq 0} H_\lambda(d)$$

and it turns out that each eigenspace $H_\lambda(d)$ is a finite dimensional vector space. In particular, we have $H_\lambda(0) = V_\lambda$.

1.3. Wess-Zumino-Witten model

Let Σ be a compact Riemann surface without boundary and fix a complex structure on Σ. We deal with smooth maps from Σ to the Lie group $G = SU(2)$. As explained in Section 1.1, the Maurer-Cartan form $\mu = X^{-1}dX$, $X \in SU(2)$, is a 1-form on $SU(2)$ with values in the Lie algebra $\mathfrak{su}(2)$ of G and is invariant under the left action of $SU(2)$ by multiplication. The 3-form

$$\sigma = \frac{1}{24\pi^2} \mathrm{Tr}(\mu \wedge \mu \wedge \mu)$$

is a left invariant volume form of $SU(2)$ and the de Rham cohomology class of σ is a generator of $H^3(SU(2), \mathbf{Z})$.

Let $f : \Sigma \to G$ be a smooth map. The integral

$$E_\Sigma(f) = -\sqrt{-1} \int_\Sigma \mathrm{Tr}(f^{-1}\partial f \wedge f^{-1}\overline{\partial} f)$$

is defined to be the *energy* of f. Hereafter, we fix an integer k called a level. The *Wess-Zumino-Witten action* $S_\Sigma(f)$ is defined by

$$S_\Sigma(f) = \frac{k}{4\pi} E_\Sigma(f) - \frac{\sqrt{-1}k}{12\pi} \int_B \mathrm{Tr}(\tilde{f}^{-1}d\tilde{f} \wedge \tilde{f}^{-1}d\tilde{f} \wedge \tilde{f}^{-1}d\tilde{f})$$

where B is a compact oriented smooth 3-manifold with boundary Σ and $\tilde{f} : B \to G$ is an extension of f over B as a smooth map. The 1-form $\tilde{f}^{-1}d\tilde{f}$ is the pull back of the Maurer-Cartan form $\mu = X^{-1}dX$ by $\tilde{f}$.

LEMMA 1.8. *For the Wess-Zumino-Witten action* $S_\Sigma(f)$

$$\exp\left(-S_\Sigma(f)\right)$$

does not depend on the choice of B *and the extension* $\tilde{f}$. *Namely,* $S_\Sigma(f)$ *depends only on the Riemann surface* Σ *and* $f : \Sigma \to G$.

PROOF. We take another compact oriented 3-manifold B' with boundary Σ and let $\tilde{f}' : \Sigma \to G$ be an extension of f over B' as a smooth map. We denote by M a 3-manifold without boundary by gluing B and $-B'$ along Σ, where $-B'$ stands for the manifold B' with the orientation reversed. Let $F : M \to G$ be a smooth map whose restriction on B and B' is $\tilde{f}$ and $\tilde{f}'$ respectively. The difference

$$\frac{k}{24\pi^2}\left(\int_B \mathrm{Tr}(\tilde{f}^{-1}d\tilde{f} \wedge \tilde{f}^{-1}d\tilde{f} \wedge \tilde{f}^{-1}d\tilde{f})\right.$$
$$\left. -\int_{B'} \mathrm{Tr}(\tilde{f}'^{-1}d\tilde{f}' \wedge \tilde{f}'^{-1}d\tilde{f}' \wedge \tilde{f}'^{-1}d\tilde{f}')\right)$$

is expressed as

$$k\int_M F^*\sigma$$

by means of the 3-form σ. Since the de Rham cohomology class of σ lies in the integral cohomology $H^3(G, \mathbf{Z})$, the value of the above integral is an integer. Therefore, $\exp\left(-S_\Sigma(f)\right)$ does not depend on the choice of B and the way of extending f over B. $\square$

The term

$$\frac{\sqrt{-1}k}{12\pi}\int_B \mathrm{Tr}(\tilde{f}^{-1}d\tilde{f} \wedge \tilde{f}^{-1}d\tilde{f} \wedge \tilde{f}^{-1}d\tilde{f})$$

in the Wess-Zumino-Witten action $S_\Sigma(f)$ is called the *Wess-Zumino term*. As we will see later, the presence of the Wess-Zumino term is essential in constructing a field theory which is invariant under local conformal transformations on a Riemann surface. Let us first explain the *Polyakov-Wiegmann formula*. For smooth maps f and g from the Riemann surface Σ to G we define their product fg pointwisely by $(fg)(z) = f(z)g(z), \ z \in \Sigma$.

PROPOSITION 1.9 (Polyakov-Wiegmann formula). *Let $f, g : \Sigma \to G$ be smooth maps. Then the equality*

$$\exp\left(-S_\Sigma(fg)\right)$$
$$= \exp\left(-S_\Sigma(f) - S_\Sigma(g) - \frac{\sqrt{-1}\,k}{2\pi} \int_\Sigma \mathrm{Tr}(f^{-1}\overline{\partial}f \wedge \partial g g^{-1})\right)$$

holds.

PROOF. Since the formula is shown by a simple calculation, the proof will only be outlined. First let us recall that for matrix valued differential forms ω of degree p and η of degree q, the equality

$$\mathrm{Tr}(\omega \wedge \eta) = (-1)^{pq}\mathrm{Tr}(\eta \wedge \omega)$$

holds. The computation of

$$I = \int_\Sigma \mathrm{Tr}((fg)^{-1}\partial(fg) \wedge (fg)^{-1}\overline{\partial}(fg))$$

with the aid of the above equality gives us $I = I_1 + I_2$ with

$$I_1 = \int_\Sigma \mathrm{Tr}(f^{-1}\partial f \wedge f^{-1}\overline{\partial}f + g^{-1}\partial g \wedge g^{-1}\overline{\partial}g),$$

$$I_2 = \int_\Sigma \mathrm{Tr}(f^{-1}\partial f \wedge \overline{\partial}g g^{-1} + \partial g g^{-1} \wedge f^{-1}\overline{\partial}f).$$

Applying $df^{-1} = -f^{-1}df f^{-1}$ and the Stokes theorem, we obtain the desired equality. □

We set

$$\Gamma_\Sigma(f, g) = -\frac{\sqrt{-1}k}{2\pi}\left(\int_\Sigma \mathrm{Tr}(f^{-1}\overline{\partial}f \wedge \partial g g^{-1})\right).$$

For the complexification $G_{\mathbf{C}} = SL(2, \mathbf{C})$ of $G = SU(2)$, consider a smooth map f_0 from the closed unit disc $D = \{z \in \mathbf{C} \;;\; |z| \leq 1\}$ to $G_{\mathbf{C}}$. The closure of the complementary space of D in the Riemann sphere $\mathbf{C}P^1 = \mathbf{C} \cup \{\infty\}$ is denoted by D_∞. That is,

$$D_\infty = \{z \in \mathbf{C} \;;\; |z| \geq 1\} \cup \{\infty\}.$$

We extend $f_0 : D \to G_{\mathbf{C}}$ as a smooth map over $\mathbf{C}P^1$ and denote it by $f : \mathbf{C}P^1 \to G_{\mathbf{C}}$. The restriction of f on D_∞ is denoted by f_∞. As in the last section, for $f : \mathbf{C}P^1 \to G_{\mathbf{C}}$ we consider

$$\exp\left(-S_{\mathbf{C}P^1}(f)\right) \in \mathbf{C}$$

where $S_{\mathbf{C}P^1}(f)$ is the Wess-Zumino-Witten action. This value is not uniquely determined by f_0 and depends on the way of extending f_0

over $\mathbf{C}P^1$. We will investigate how it depends on the choice of an extension.

We take another extension of $f_0 : D \to G_{\mathbf{C}}$ over $\mathbf{C}P^1$ and denote it by $f' : \mathbf{C}P^1 \to G_{\mathbf{C}}$. We have a smooth map $h : \mathbf{C}P^1 \to G_{\mathbf{C}}$ with $f' = fh$. Here $h(z)$ is the unit element e of $G_{\mathbf{C}}$ for any $z \in D$. The restriction of h on D_∞ is denoted by h_∞. By the Polyakov-Wiegmann formula

$$\exp\left(-S_{\mathbf{C}P^1}(fh)\right) = \exp\left(-S_{\mathbf{C}P^1}(f) - S_{\mathbf{C}P^1}(h) + \Gamma_{\mathbf{C}P^1}(f,h)\right)$$

holds. Moreover, since $h(z) = e$ for any $z \in D$ we have

$$\Gamma_{\mathbf{C}P^1}(f,h) = \Gamma_{D_\infty}(f_\infty, h_\infty).$$

Hence we obtain the equation

(1.12)
$$\exp\left(-S_{\mathbf{C}P^1}(fh)\right) = \exp\left(-S_{\mathbf{C}P^1}(f) - S_{\mathbf{C}P^1}(h) + \Gamma_{D_\infty}(f_\infty, h_\infty)\right).$$

Based on the above observation we construct a complex line bundle over $LG_{\mathbf{C}}$ and explain that it is natural to consider $\exp\left(-S_D(f_0)\right)$ as an element of a fibre of this complex line bundle. Denote by $\mathrm{Map}_0(D_\infty, G_{\mathbf{C}})$ the set of the smooth maps $\varphi : D_\infty \to G_{\mathbf{C}}$ with $\varphi(\infty) = e$. By means of the polar coordinates we write $z^{-1} = re^{\sqrt{-1}\theta}$ and set $\varphi(re^{\sqrt{-1}\theta}) = p_r(e^{\sqrt{-1}\theta})$. The map

$$p_r : S^1 \to G_{\mathbf{C}}, \quad 0 \le r \le 1$$

defines a loop of $G_{\mathbf{C}}$ for any fixed r with $0 \le r \le 1$. In particular, when $r = 0$, the above map p_0 is a constant loop sending any point in S^1 to the unit element e of $G_{\mathbf{C}}$. The one-parameter family of loops $p_r, 0 \le r \le 1$, in $G_{\mathbf{C}}$ corresponds to a path in the loop group $LG_{\mathbf{C}}$ starting at the unit element e of $LG_{\mathbf{C}}$. Thus, the set $\mathrm{Map}_0(D_\infty, G_{\mathbf{C}})$ is in one-to-one correspondence with the set of smooth paths in $LG_{\mathbf{C}}$ starting at e.

We introduce the following equivalence relation $\sim$ on the Cartesian product

$$\mathrm{Map}_0(D_\infty, G_{\mathbf{C}}) \times \mathbf{C}.$$

For $(f_\infty, u), (g_\infty, v) \in \mathrm{Map}_0(D_\infty, G_{\mathbf{C}})$ we set

$$(f_\infty, u) \sim (g_\infty, v)$$

if and only if the following two conditions are satisfied.

1. $f_\infty(z) = g_\infty(z)$ holds for any $z \in \partial D$.

2. Expressing g_∞ as $g_\infty = f_\infty h_\infty$ one has

$$v = u \exp\left(-S_{\mathbf{C}P^1}(h) + \Gamma_{D_\infty}(f_\infty, h_\infty)\right)$$

where $h : \mathbf{C}P^1 \to G_{\mathbf{C}}$ is the extension of $h_\infty : D_\infty \to G_{\mathbf{C}}$ over $\mathbf{C}P^1$ such that $h(z) = e$ holds for any $z \in D$.

The set of equivalence classes

$$\mathrm{Map}_0(D_\infty, G_{\mathbf{C}}) \times \mathbf{C}/ \sim$$

by the above equivalence relation $\sim$ is denoted by $\mathcal{L}$. We define the projection map $\pi : \mathcal{L} \to LG_{\mathbf{C}}$ by

$$\pi([f_\infty, u]) = f_\infty \circ \iota$$

where $[f_\infty, u]$ is the equivalence class represented by (f_∞, u) and $\iota : \partial D \to D$ is the inclusion map. It turns out that by the above projection map π the set of equivalence classes $\mathcal{L}$ has a structure of a complex line bundle over $LG_{\mathbf{C}}$. As explained above, each element in $\mathrm{Map}_0(D_\infty, G_{\mathbf{C}}) \times \mathbf{C}$ corresponds to a path in $LG_{\mathbf{C}}$. We denote by γ_1 and γ_2 the paths corresponding to f_∞ and g_∞ respectively. The paths

$$\gamma_i : [0,1] \to LG_{\mathbf{C}}, \ i = 1, 2,$$

satisfy $\gamma_1(0) = \gamma_2(0) = e$ and the above condition 1 implies $\gamma_1(1) = \gamma_2(1)$. The condition 2 measures the holonomy along the loop $\gamma_1 \cdot \gamma_2^{-1}$. We have a connection on $\mathcal{L}$ which gives the above holonomy. The construction of the complex line bundle $\mathcal{L}$ over the loop group $LG_{\mathbf{C}}$ is analogous to that in Section 1.1. It can be shown that the complex line bundle $\mathcal{L}$ is topologically isomorphic to the k-fold tensor product of the fundamental line bundle in Section 1.1.

We take an extension $f : \mathbf{C}P^1 \to G_{\mathbf{C}}$ of $f_0 : D \to G_{\mathbf{C}}$ and define $\exp\left(-S_D(f_0)\right)$ as the equivalence class $[f_\infty, \exp(-S_{\mathbf{C}P^1}(f))]$.

LEMMA 1.10. *For $f_0 : D \to G_{\mathbf{C}}$, the equivalence class*

$$[f_\infty, \exp(-S_{\mathbf{C}P^1}(f))]$$

does not depend on the choice of an extension of f_0 on $\mathbf{C}P^1$. Therefore it defines an element in the fibre of the line bundle $\mathcal{L}$ over $f_0 \circ \iota$.

PROOF. We take another extension $f' : \mathbf{C}P^1 \to G_{\mathbf{C}}$ of $f_0 : D \to G_{\mathbf{C}}$ and write it as $f' = fh$. It is enough to show that

$$(f_\infty, \exp(-S_{\mathbf{C}P^1}(f))) \sim (f_\infty h_\infty, \exp(-S_{\mathbf{C}P^1}(fh))),$$

but this equivalence relation follows immediately from the equation (1.12). $\qquad \square$

Let us construct the dual line bundle $\mathcal{L}^{-1}$ of $\mathcal{L}$. Denote by $\mathrm{Map}_0(D, G_{\mathbf{C}})$ the set of smooth maps $\varphi : D \to G_{\mathbf{C}}$ with $\varphi(0) = e$ and define an equivalence relation on $\mathrm{Map}_0(D, G_{\mathbf{C}}) \times \mathbf{C}$ in the following way. For $(f_0, u), (g_0, v) \in \mathrm{Map}_0(D, G_{\mathbf{C}}) \times \mathbf{C}$ we set

$$(f_0, u) \sim (g_0, v)$$

if and only if the conditions below are satisfied.

1. $f_0(z) = g_0(z)$ holds for any $z \in \partial D$.
2. Writing $g_0 = f_0 h_0$ one has

$$v = u \exp\left(-S_{\mathbf{CP}^1}(h) + \Gamma_D(f_0, h_0)\right).$$

We define $\mathcal{L}^{-1}$ as the set of equivalence classes

$$\mathrm{Map}_0(D, G_{\mathbf{C}}) \times \mathbf{C}/\sim .$$

Then $\mathcal{L}^{-1}$ has a structure of a complex line bundle over the loop group $LG_{\mathbf{C}}$ with the projection map π defined by $\pi([f_0, u]) = f_0 \circ \iota$. As in Lemma 1.10 we see that for $f_\infty : D_\infty \to G_{\mathbf{C}}$,

$$\exp\left(-S_{D_\infty}(f_\infty)\right) = [f_0, \exp\left(-S_{\mathbf{CP}^1}(f)\right)]$$

is well defined as an element of the fibre of $\mathcal{L}^{-1}$ over $f_\infty \circ \iota$.

Suppose that smooth maps $f_0 : D \to G_{\mathbf{C}}$ and $f_\infty : D \to G_{\mathbf{C}}$ such that $f_0 \circ \iota = f_\infty \circ \iota$ are given. We have a map $f : \mathbf{C}P^1 \to G_{\mathbf{C}}$ whose restriction on D and D_∞ are f_0 and f_∞ respectively. We denote by $\gamma : S^1 \to G_{\mathbf{C}}$ the loop defined by $f \circ \iota$. We have a pairing

$$\mathcal{L}_\gamma \times \mathcal{L}_\gamma^{-1} \to \mathbf{C}$$

defined by

$$\langle [f_\infty, u], [f_0, v] \rangle = uv \exp\left(S_{\mathbf{CP}^1}(f)\right)$$

where $\mathcal{L}_\gamma$ is the fibre of $\mathcal{L}$ over γ. It follows from the equation (1.12) that the right hand side of the above equation does not depend on the choice of representatives of the equivalence classes and the pairing is well defined. It follows directly from the definition that with respect to the above pairing

$$\langle \exp\left(-S_D(f_0)\right), \exp\left(-S_{D_\infty}(f_\infty)\right) \rangle = \exp\left(-S_{\mathbf{CP}^1}(f)\right)$$

holds.

The complex line bundle $\pi : \mathcal{L} \to LG_{\mathbf{C}}$ leads us to a central extension of the loop group $LG_{\mathbf{C}}$. The map $s : LG_{\mathbf{C}} \to \mathcal{L}$ with $s(\gamma) = 0 \in \mathcal{L}_\gamma$ for any $\gamma \in LG_{\mathbf{C}}$ is called the zero section of $\mathcal{L}$. We set

$$\widehat{LG_{\mathbf{C}}} = \mathcal{L} \setminus s(LG_{\mathbf{C}})$$

for the zero section s and we will define a group structure on $\widehat{LG_{\mathbf{C}}}$. Let us notice that any element of the fibre $\mathcal{L}_\gamma$ of the complex line bundle $\mathcal{L}$ over a loop $\gamma : S^1 \to G_{\mathbf{C}}$ can be written as

$$z \exp\left(-S_D(g)\right), \quad z \in \mathbf{C},$$

where $g : D \to G_{\mathbf{C}}$ is a smooth map such that $\gamma = g \circ \iota$. For $g_i : D \to G_{\mathbf{C}}$, $i = 1, 2$, we put $\gamma_i = g \circ \iota$ and we define the product of $\exp\left(-S_D(g_1)\right) \in \mathcal{L}_{\gamma_1}$ and $\exp\left(-S_D(g_2)\right) \in \mathcal{L}_{\gamma_2}$ by

(1.13)
$$\exp\left(-S_D(g_1)\right) \bullet \exp\left(-S_D(g_2)\right) = \exp\left(-\Gamma_D(g_1, g_2)\right)\exp\left(-S_D(g_1 g_2)\right).$$

Extending this as a complex linear map we obtain the product

$$\mathcal{L}_{\gamma_1} \times \mathcal{L}_{\gamma_2} \to \mathcal{L}_{\gamma_1 \cdot \gamma_2}.$$

It can be shown that the above product does not depend on the choice of g_1 and g_2. Moreover, the associativity of the product follows from the equality

$$\Gamma_D(g_2, g_3) + \Gamma_D(g_1 g_2, g_3) - \Gamma_D(g_1, g_2 g_3) + \Gamma_D(g_1, g_2) = 0$$

for smooth maps $g_i : D \to G_{\mathbf{C}}$, $i = 1, 2, 3$. The unit element with respect to the above multiplication is $\exp\left(-S_D(e)\right)$ where $e : D \to G_{\mathbf{C}}$ is the map such that $e(z)$ is the unit element of $G_{\mathbf{C}}$ for any $z \in D$. Thus $\widehat{LG_{\mathbf{C}}}$ is equipped with a structure of a group. We define $i : \mathbf{C}^* \to \widehat{LG_{\mathbf{C}}}$ by $i(z) = z \exp\left(-S_D(e)\right)$. We obtain an exact sequence of groups

$$1 \to \mathbf{C}^* \overset{\iota}{\to} \widehat{LG_{\mathbf{C}}} \overset{\pi}{\to} LG_{\mathbf{C}} \to 1,$$

which gives a central extension of the loop group $LG_{\mathbf{C}}$.

Let Σ be a compact Riemann surface with boundary and fix a complex structure on Σ. The boundary $\partial\Sigma$ is homeomorphic to the disjoint union of finitely many circles. We fix a diffeomorphism $p_i : S^1 \to \partial\Sigma$, $1 \le i \le m$, for each connected component of $\partial\Sigma$. We have $\partial\Sigma = \bigcup_{i=1}^m p_i(S^1)$ and the orientation of $\partial\Sigma$ is the one induced from that of Σ. As depicted in Figure 1.1, we glue the boundary of unit discs D_i, $1 \le i \le m$, with $p_i(S^1)$, $1 \le i \le m$, to obtain a closed Riemann surface $\widetilde{\Sigma}$.

Now we are in position to define $\exp\left(-S_\Sigma(g)\right)$ for a smooth map $g : \Sigma \to G_{\mathbf{C}}$. We extend g as a smooth map over $\widetilde{\Sigma}$ and denote it by $\tilde{g} : \widetilde{\Sigma} \to G_{\mathbf{C}}$. The restriction of g on D_i is denoted by g_i. Applying the above construction of the complex line bundle $\mathcal{L}^{-1}$ to each unit disc D_i, we can define $\exp\left(-S_{D_i}(g_i)\right)$ as an element of the fibre $\mathcal{L}_{g \circ p_i}^{-1}$ of

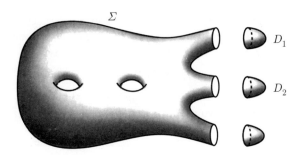

FIGURE 1.1. Attaching discs $D_1, D_2, \cdots$ to the boundary of Σ

the complex line bundle $\mathcal{L}^{-1}$. Based on this, we define $\exp\left(-S_\Sigma(g)\right)$ as the element of $\bigotimes_{i=1}^m \mathcal{L}^{-1}_{g \circ p_i}$ specified by

$$\langle \exp\left(-S_\Sigma(g)\right), \bigotimes_{i=1}^m \exp\left(-S_{D_i}(g_i)\right)\rangle = \exp\left(-S_{\tilde{\Sigma}}(\tilde{g})\right).$$

Again by the Polyakov-Wiegmann formula one can verify that it does not depend on the choice of an extension $\tilde{g}$.

Thus for a smooth map $g : \Sigma \to G_{\mathbf{C}}$ we have defined $\exp\left(-S_\Sigma(g)\right)$ as an element of the tensor product $\bigotimes_{i=1}^m \mathcal{L}_{g \circ p_i}$ of fibres of the complex line bundle $\mathcal{L}$. In the following we deal with the case $m = 1$ for simplicity. We fix a diffeomorphism $p : S^1 \to \partial\Sigma$. For a smooth map $f : \Sigma \to G_{\mathbf{C}}$ we define the left action $l(f)$ on $\mathcal{L}$ by

$$(1.14) \qquad l(f)\exp\left(-S_\Sigma(g)\right) = \exp\left(-S_\Sigma(f)\right) \bullet \exp\left(-S_\Sigma(g)\right)$$

using the product defined in the equation (1.13). It follows from the Polyakov-Wiegmann formula that the consequence of this action is also expressed as

$$(1.15) \quad \exp\left(-S_\Sigma(f)\right) \bullet \exp\left(-S_\Sigma(g)\right) = \exp\left(-S_\Sigma(fg) - \Gamma_\Sigma(f, g)\right).$$

Similarly, the right action $r(f)$ on $\mathcal{L}$ is defined by

$$(1.16) \qquad r(f)\exp\left(-S_\Sigma(g)\right) = \exp\left(-S_\Sigma(g)\right) \bullet \exp\left(-S_\Sigma(f)\right).$$

The set $\mathrm{Map}(\Sigma, G_{\mathbf{C}})$ consisting of smooth maps $f : \Sigma \to G_{\mathbf{C}}$ has a structure of a group by the pointwise multiplication. It is identified with the gauge group of a topologically trivial principal $G_{\mathbf{C}}$ bundle over Σ. The composition $\iota \circ p$ of a diffeomorphism $p : S^1 \to \partial\Sigma$ and the inclusion $\iota : \partial\Sigma \to \Sigma$ induces the restriction map

$$(\iota \circ p)^* : \mathrm{Map}(\Sigma, G_{\mathbf{C}}) \to LG_{\mathbf{C}}.$$

Here $(\iota \circ p)^* f = f \circ \iota \circ p$. In the following, we write $f|_{\partial \Sigma}$ for $(\iota \circ p)^* f$. We have a left action and a right action of the gauge group $\mathrm{Map}(\Sigma, G_{\mathbf{C}})$ on the loop group $LG_{\mathbf{C}}$ by the left multiplication and by the right multiplication respectively via the above restriction map $(\iota \circ p)^*$. This action of the gauge group $\mathrm{Map}(\Sigma, G_{\mathbf{C}})$ can be lifted to the total space of the line bundle $\pi : \mathcal{L} \to LG_{\mathbf{C}}$ by means of the action defined in the equations (1.14) and (1.16). Namely, for $f \in \mathrm{Map}(\Sigma, G_{\mathbf{C}})$, the left action $l(f)$ on $\mathcal{L}$ satisfies

$$\pi\left(l(f)x\right) = \left(f|_{\partial\Sigma}\right) \cdot \pi(x), \ x \in \mathcal{L},$$

and the right action $r(f)$ on $\mathcal{L}$ satisfies

$$\pi\left(r(f)x\right) = \pi(x) \cdot \left(f|_{\partial\Sigma}\right), \ x \in \mathcal{L}.$$

The above lifting permits us to define a representation of $\mathrm{Map}(\Sigma, G_{\mathbf{C}})$ on the space of sections $\Gamma(\mathcal{L})$. We define a representation $\rho : \mathrm{Map}(\Sigma, G_{\mathbf{C}}) \to \mathrm{Aut}(\Gamma(\mathcal{L}))$ by

$$[\rho(f)s](\gamma) = l(f) \, s\left(\left(f|_{\partial\Sigma}\right)^{-1} \cdot \gamma\right), \ s \in \Gamma(\mathcal{L}), \gamma \in LG_{\mathbf{C}},$$

for $f \in \mathrm{Map}(\Sigma, G_{\mathbf{C}})$. Similarly, using the right action $r(f)$ on $\mathcal{L}$ we define a representation $\rho^* : \mathrm{Map}(\Sigma, G_{\mathbf{C}}) \to \mathrm{Aut}(\Gamma(\mathcal{L}))$ by

$$[\rho^*(f)s](\gamma) = r(f^*) \, s\left(\gamma \cdot \left(f^*|_{\partial\Sigma}\right)^{-1}\right), \ s \in \Gamma(\mathcal{L}), \gamma \in LG_{\mathbf{C}},$$

where $f^*(z) = {}^t\overline{f(z)}$.

Let $h : \Sigma \to G_{\mathbf{C}}$ be a smooth map which is holomorphic on the set of interior points $\mathrm{Int}\,\Sigma$ of Σ. The map $h^* : \Sigma \to G_{\mathbf{C}}$ is anti-holomorphic on $\mathrm{Int}\,\Sigma$. The left action $l(h)$ and the right action $r(h^*)$ on $\mathcal{L}$ behave in a covariant way in the following sense.

PROPOSITION 1.11. *Let* $g : \Sigma \to G_{\mathbf{C}}$ *be a smooth map and let* $h : \Sigma \to G_{\mathbf{C}}$ *be a smooth map which is holomorphic on* $\mathrm{Int}\,\Sigma$. *Then, we have*

$$l(h)\exp\left(-S_\Sigma(g)\right) = \exp\left(-S_\Sigma(hg)\right)$$

and for the anti-holomorphic map $h^* : \Sigma \to G_{\mathbf{C}}$ *we have*

$$r(h^*)\exp\left(-S_\Sigma(g)\right) = \exp\left(-S_\Sigma(gh^*)\right).$$

PROOF. Since $\overline{\partial} h = 0$ holds for a holomorphic map h we have

$$\Gamma_\Sigma(h, g) = -\frac{\sqrt{-1}k}{2\pi}\left(\int_\Sigma \mathrm{Tr}(h^{-1}\overline{\partial}h \wedge \partial g g^{-1})\right) = 0.$$

Combining with the Polyakov-Wiegmann formula in the form of the equation (1.15), we obtain the desired result. The equality for an anti-holomorphic map h^* follows from $\partial h^* = 0$ in a similar way. $\square$

Let us describe the representations of the gauge group $\text{Map}(\Sigma, G_{\mathbf{C}})$ on the space of sections $\Gamma(\mathcal{L})$ at the infinitesimal level. We consider the case when Σ is the unit disc D. For a non-negative integer n and $X \in \mathfrak{g}$ we put

$$X_{n,\varepsilon}(z) = e^{\varepsilon X z^n}, \quad z \in D, \varepsilon \in \mathbf{R},$$

and for a negative integer n we put

$$X_{n,\varepsilon}(z) = e^{\varepsilon X \bar{z}^{-n}}, \quad z \in D, \varepsilon \in \mathbf{R}.$$

Then, $X_{n,\varepsilon}$ is a map from the unit disc D to the Lie group $G_{\mathbf{C}}$. The infinitesimal version of the action of $X_{n,\varepsilon}$ by ρ is defined by

$$X_n s = \left. \frac{d}{d\varepsilon} \right|_{\varepsilon=0} \rho(X_{n,\varepsilon}) s, \quad s \in \Gamma(\mathcal{L}).$$

We have the following lemma.

LEMMA 1.12. *The operators X_m and Y_n, $m, n \in \mathbf{Z}$, satisfy the relation*

$$[X_m, Y_n] = [X, Y]_{m+n} + mk\delta_{m+n,0}\langle X, Y \rangle.$$

PROOF. We put $f = X_{m,\varepsilon_1}$ and $g = Y_{n,\varepsilon_2}$ for $\varepsilon_1, \varepsilon_2 \in \mathbf{R}$. In the case $m, n \geq 0$ or $m, n \leq 0$ the relation $[X_m, Y_n] = [X, Y]_{m+n}$ follows immediately from

$$\Gamma_D(f, g) = \Gamma_D(g, f) = 0.$$

Let us suppose $m \geq 0$ and $n \leq 0$. We have $\Gamma_D(f, g) = 0$. For $\Gamma_D(g, f)$ we have

$$\lim_{\varepsilon_1, \varepsilon_2 \to 0} \frac{1}{\varepsilon_1 \varepsilon_2} \Gamma_D(g, f) = \frac{k}{2\pi\sqrt{-1}} \int_D \text{Tr}\left(mz^{m-1} X \, dz \wedge n\bar{z}^{-n-1} Y \, d\bar{z} \right),$$

which is equal to $mk\langle X, Y \rangle$ if $m = -n$ and is equal to zero otherwise. This settles the case $m \geq 0$, $n \leq 0$. One can perform a similar computation in the case $m \leq 0$, $n \geq 0$. $\qquad \square$

The above lemma shows that the map defined by the correspondence $X \otimes t^n \mapsto X_n$ gives a representation of the affine Lie algebra $\widehat{\mathfrak{g}}$ on the space of sections $\Gamma(\mathcal{L})$. The central element c in $\widehat{\mathfrak{g}}$ acts as the scalar multiplication by k. In a similar way one can construct a representation of $\widehat{\mathfrak{g}}$ on $\Gamma(\mathcal{L})$ based on the representation ρ^*. We define the operator $\overline{X}_n$ acting on $\Gamma(\mathcal{L})$ by

$$\overline{X}_n s = \left. \frac{d}{d\varepsilon} \right|_{\varepsilon=0} \rho^*(X_{n,\varepsilon}) s, \quad s \in \Gamma(\mathcal{L}).$$

The correspondence $X \otimes t^n \mapsto \overline{X}_n$ also defines a representation of $\widehat{\mathfrak{g}}$. In other words, the operators $\overline{X}_m$ and $\overline{Y}_n$, $m, n \in \mathbf{Z}$, satisfy the relation

$$[\overline{X}_m, \overline{Y}_n] = \overline{[X, Y]}_{m+n} + mk\delta_{m+n,0}\langle X, Y\rangle.$$

Notice that we have

$$[X_m, \overline{Y}_n] = 0$$

for any $X, Y \in \widehat{\mathfrak{g}}$ and $m, n \in \mathbf{Z}$. Thus, the space of sections $\Gamma(\mathcal{L})$ carries two types of representations of the affine Lie algebra $\widehat{\mathfrak{g}}$ based on ρ and ρ^* respectively. These two representations commute with each other.

A smooth section $\varphi \in \Gamma(\mathcal{L})$ is called *primary* if and only if

$$X_n\varphi = \overline{X}_n\varphi = 0$$

holds for any $X \in \mathfrak{g}$ and $n > 0$. We identify $G_{\mathbf{C}}$ with the subset of $LG_{\mathbf{C}}$ consisting of constant maps. With this identification we write

$$\varphi(g) = \varphi_0(g)\exp\left(-S_D(g)\right)$$

for $g \in G_{\mathbf{C}}$. Here φ_0 is a complex valued function on $G_{\mathbf{C}}$ and in the expression $S_D(g)$ we regard g as a constant map on D with value g. We set $\operatorname{Int} D = \{z \in \mathbf{C} \mid |z| < 1\}$ as before.

LEMMA 1.13. *Let* $g_i : D \to G_{\mathbf{C}}$, $i = 1, 2$, *be smooth maps with* $g_i(0) = e$ *which are holomorphic on* $\operatorname{Int} D$. *If* $\varphi \in \Gamma(\mathcal{L})$ *is primary, then we have*

$$\varphi((g_1gg_2^*)|_{\partial D}) = \varphi_0(g)\exp\left(-S_D(g_1gg_2^*)\right)$$

for any $g \in G_{\mathbf{C}}$.

PROOF. Since φ is primary it follows from the definition that

$$\rho(g_1)\rho^*(g_2)\varphi = \varphi$$

holds. Hence we obtain

$$\varphi((g_1gg_2^*)|_{\partial D}) = l(g_1)r(g_2^*)\varphi(g).$$

Applying Proposition 1.11, we have

$$l(g_1)r(g_2^*)\varphi(g) = \varphi_0(g)\exp\left(-S_D(g_1gg_2^*)\right).$$

This completes the proof.   □

We denote by $L^+G_{\mathbf{C}}$ the subgroup of the loop group $LG_{\mathbf{C}}$ consisting of loops which are boundary values of holomorphic maps

$$f : \operatorname{Int} D \to G_{\mathbf{C}}.$$

Similarly, we denote by $L^-G_{\mathbf{C}}$ the subgroup of loops which are boundary values of anti-holomorphic maps

$$f^* : \operatorname{Int} D \to G_{\mathbf{C}}.$$

To construct highest weight representations of $\widehat{\mathfrak{g}}$ in $\Gamma(\mathcal{L})$ the following theorem due to Birkhoff plays an important role.

THEOREM 1.14 (Birkhoff). *Any loop $\gamma \in LG_{\mathbf{C}}$ can be factorized as*

$$\gamma = \gamma_+ \cdot \lambda \cdot \gamma_-$$

where $\gamma_+ \in L^+G_{\mathbf{C}}$, $\gamma_- \in L^-G_{\mathbf{C}}$ and λ is a homomorphism from S^1 into the diagonal matrices in $G_{\mathbf{C}}$. Loops for which λ is the identity matrix form a dense open subset of $LG_{\mathbf{C}}$.

We refer the reader to [**46**] for the proof of the Birkhoff theorem. Let us consider the subgroup $L_0^+G_{\mathbf{C}}$ of $L^+G_{\mathbf{C}}$ consisting of loops γ which are boundary values of holomorphic maps $f : \operatorname{Int} D \to G_{\mathbf{C}}$ with $f(0) = e$. We define $L_0^-G_{\mathbf{C}}$ in the same way as boundary values of anti-holomorphic maps $f^* : \operatorname{Int} D \to G_{\mathbf{C}}$ with $f^*(0) = e$. It follows from the Birkhoff theorem that the image of the multiplication map

$$(1.17) \qquad L_0^+G_{\mathbf{C}} \times G_{\mathbf{C}} \times L_0^-G_{\mathbf{C}} \to LG_{\mathbf{C}}$$

is a dense open subset of $LG_{\mathbf{C}}$. Here we identify $G_{\mathbf{C}}$ with the set of constant loops. Combining with Lemma 1.13, we conclude that a primary section $\varphi \in \Gamma(\mathcal{L})$ is uniquely determined by its associated function φ_0 on $G_{\mathbf{C}}$.

Let $r : G_{\mathbf{C}} \to \operatorname{Aut}(V_\lambda)$ be a representation of $G_{\mathbf{C}}$ whose infinitesimal action of $\mathfrak{g}$ is isomorphic to the irreducible representation with highest weight $\lambda \in \mathbf{Z}$. For a linear transformation α of V_λ consider the function χ_α on $G_{\mathbf{C}}$ defined by

$$\chi_\alpha(g) = \operatorname{Tr}(\alpha \circ r(g)), \ g \in G_{\mathbf{C}}.$$

We fix r and we define W_λ to be the set of functions on $G_{\mathbf{C}}$ which can be expressed as χ_α with some $\alpha \in \operatorname{End}(V_\lambda)$. The Lie group $G_{\mathbf{C}}$ acts on W_λ by

$$(g \cdot \chi_\alpha)(x) = \chi_\alpha(gx) = \chi_{\alpha \circ r(g)}(x), \ g, x \in G_{\mathbf{C}}.$$

As the infinitesimal version of the above action, W_λ can be regarded as a $\mathfrak{g}$ module and we have an isomorphism

$$W_\lambda \cong V_\lambda \otimes V_\lambda^*$$

where V_λ is an irreducible $\mathfrak{g}$ module with highest weight λ and V_λ^* is its dual module.

For $\varphi_0 \in W_\lambda$ we define a section φ of the line bundle $\mathcal{L}$ over the image of the multiplication map (1.17) by

$$(1.18) \quad \varphi(g_1 g g_2^*) = \varphi_0(g) \exp\left(-S_D(g_1 g g_2^*)\right),$$

$$g \in G_{\mathbf{C}}, g_1 \in L_0^+ G_{\mathbf{C}}, g_2 \in L_0^- G_{\mathbf{C}}.$$

Let us suppose that k is a positive integer. We have the following proposition.

PROPOSITION 1.15. *If the highest weight λ satisfies $0 \leq \lambda \leq k$, then for $\varphi_0 \in W_\lambda$ the section φ defined by (1.18) extends over $LG_{\mathbf{C}}$. The set of all sections φ with $\varphi_0 \in W_\lambda$ admits an action of $\widehat{\mathfrak{g}}$ and is isomorphic to the tensor product of the integrable highest weight modules $H_\lambda \otimes H_\lambda^*$ as a $\widehat{\mathfrak{g}}$ module.*

We give a sketch of the proof of the above proposition. See [24] for details. According to Birkhoff's theorem, we have a stratification of the loop group $LG_{\mathbf{C}}$ by

$$\mathcal{S}_n = \{\gamma_+ \cdot \lambda \cdot \gamma_- \mid \gamma_+ \in L^+ G_{\mathbf{C}},\ \gamma_- \in L^- G_{\mathbf{C}},\ \lambda(z) = \mathrm{diag}(z^n, z^{-n})\}$$

where n is a non-negative integer. Namely, we have a decomposition of $LG_{\mathbf{C}}$ into the disjoint union

$$LG_{\mathbf{C}} = \bigcup_{n \geq 0} \mathcal{S}_n,$$

and $\mathcal{S}_0$ is a dense open set. It follows from Lemma 1.13 and Birkhoff's theorem that $\varphi_0 \in W_\lambda$ is extended over $\mathcal{S}_0$ as a primary field φ defined by (1.18). The point is that φ is extended to $LG_{\mathbf{C}}$ if the condition $0 \leq k \leq \lambda$ is satisfied. To show this we look at the singularities of φ along the codimension one stratum $\mathcal{S}_1$. The primary field φ plays the role of the highest weight vector and we see that as a module over $\widehat{\mathfrak{g}}$, the primary field φ generates the integrable highest weights module $H_\lambda \otimes H_\lambda^*$.

Proposition 1.15 shows that the space of the sections of the complex line bundle $\mathcal{L}$ contains as a subspace the direct sum

$$\bigoplus_{0 \leq \lambda \leq k} H_\lambda \otimes H_\lambda^*.$$

Let us now recall the original motivation of the Wess-Zumino-Witten model. For a closed Riemann surface $\tilde{\Sigma}$ we consider the partition function, i.e., the "average" of $\exp(-S_{\tilde{\Sigma}}(f))$ for all smooth maps $f :$ $\tilde{\Sigma} \to G_{\mathbf{C}}$. Formally it is expressed as Feynman's path integral

$$\int_{f:\tilde{\Sigma}\to G_{\mathbf{C}}} \exp(-S_{\tilde{\Sigma}}(f)) \mathcal{D}f$$

but this integral over the infinite dimensional space of smooth maps from $\tilde{\Sigma}$ to $G_{\mathbf{C}}$ has not been established rigorously. Our strategy is to consider the case of a Riemann surface with boundary. Let Σ be a Riemann surface with boundary homeomorphic to S^1 and let $\iota : S^1 \to \Sigma$ be the inclusion map. We have shown that $\exp(-S_{\Sigma}(f))$ is regarded as an element of the fibre over $f \circ \iota$ of the complex line bundle $\mathcal{L}$ over $LG_{\mathbf{C}}$. We fix an element γ of the loop group $LG_{\mathbf{C}}$. Let us consider heuristically the above Feynman integral over the space of smooth maps $f : \Sigma \to G_{\mathbf{C}}$ such that $f \circ \iota$ coincides with γ. The result should be a section of the line bundle $\mathcal{L}$. Instead of giving a precise meaning to this integral we determine this section by its expected properties. It follows from Proposition 1.11 that this section should be invariant under the left action of a holomorphic map $h : \Sigma \to G_{\mathbf{C}}$ and the right action of an anti-holomorphic map $h^* : \Sigma \to G_{\mathbf{C}}$. Our section will be regarded as an element of $\oplus_{0\le\lambda\le k}H_\lambda \otimes H_\lambda^*$ with the above invariance property under holomorphic and anti-holomorphic maps. Finally, for a closed Riemann surface we decompose it into two parts with common boundary S^1 and the partition function will be obtained by means of the pairing $\mathcal{L} \times \mathcal{L}^{-1} \to \mathbf{C}$. This motivated the algebraic construction in the next section.

1.4. The space of conformal blocks and fusion rules

We deal with the Riemann sphere $\mathbf{C}P^1$ with a homogeneous coordinate $[\zeta_0 : \zeta_1]$. We identify $\mathbf{C}P^1$ with the one point compactification, $\mathbf{C} \cup \{\infty\}$, of the complex plane $\mathbf{C}$. Here $z = \zeta_0/\zeta_1$ is considered to be a coordinate function for the complex plane $\mathbf{C}$. Take n distinct points $p_1, \cdots, p_n$ on $\mathbf{C}P^1$ and for $1 \le j \le n$ we denote by z_j the coordinate $z(p_j)$ of the point p_j. In the case $p_j \ne \infty$ we put $t_j = z - z_j$. If p_j coincides with ∞, then we put $t_j = 1/z$. We see that t_j gives a local coordinate function around p_j with $t_j(p_j) = 0$. We fix this coordinate function t_j around each p_j, $1 \le j \le n$. In the following, we suppose for simplicity that none of the p_j, $1 \le j \le n$, coincides with ∞.

We denote by $\mathcal{M}_{p_1,\cdots,p_n}$ the vector space of meromorphic functions on $\mathbf{C}P^1$ with poles of any order at most at $p_1,\cdots,p_n$. We set

$$\mathfrak{g}(p_1,\cdots,p_n) = \mathfrak{g} \otimes \mathcal{M}_{p_1,\cdots,p_n}$$

where $\mathfrak{g}$ is the Lie algebra $sl_2(\mathbf{C})$ as in the last section. An element of $\mathfrak{g}(p_1,\cdots,p_n)$ is a meromorphic function on $\mathbf{C}P^1$ with values in $\mathfrak{g}$ which has poles at most at $p_1,\cdots,p_n$. The above $\mathfrak{g}(p_1,\cdots,p_n)$ has a structure of a Lie algebra with the Lie bracket defined by

$$[X \otimes f, Y \otimes g] = [X,Y] \otimes fg, \quad X,Y \in \mathfrak{g}, \quad f,g \in \mathcal{M}_{p_1,\cdots,p_n}.$$

By the Laurent expansion of an element of $\mathfrak{g}(p_1,\cdots,p_n)$ at p_j with respect to the coordinate function t_j we obtain a linear map

$$\tau_j : \mathfrak{g}(p_1,\cdots,p_n) \to \mathfrak{g} \otimes \mathbf{C}((t_j))$$

for each j, $1 \le j \le n$. Composing with a natural injection from $\mathfrak{g} \otimes \mathbf{C}((t_j))$ to the affine Lie algebra $\widehat{\mathfrak{g}}_j = \mathfrak{g} \otimes \mathbf{C}((t_j)) \oplus \mathbf{C}c$, we have a linear map

$$\iota_j : \mathfrak{g}(p_1,\cdots,p_n) \to \widehat{\mathfrak{g}}_j.$$

In the following, we fix a positive integer k called a level. To each point p_j, $1 \le j \le n$, we associate the integrable highest weight $\widehat{\mathfrak{g}}_j$ module H_{λ_j} with highest weight λ_j. We define the diagonal action Δ of $\mathfrak{g}(p_1,\cdots,p_n)$ on the tensor product $H_{\lambda_1} \otimes \cdots \otimes H_{\lambda_n}$ by

$$\Delta(\varphi)(\xi_1 \otimes \cdots \otimes \xi_n)$$
$$= \sum_{j=1}^n \xi_1 \otimes \cdots \otimes \iota_j(\varphi)\xi_j \otimes \cdots \otimes \xi_n$$

for $\varphi \in \mathfrak{g}(p_1,\cdots,p_n)$ and $\xi_j \in H_{\lambda_j}$, $1 \le j \le n$.

LEMMA 1.16. *The above action*

$$\Delta : \mathfrak{g}(p_1,\cdots,p_n) \to \mathrm{End}(H_{\lambda_1} \otimes \cdots \otimes H_{\lambda_n})$$

is a representation of the Lie algebra $\mathfrak{g}(p_1,\cdots,p_n)$.

PROOF. For $f \in \mathcal{M}_{p_1,\cdots,p_n}$ we denote by f_{p_j} the Laurent series in t_j obtained as the Laurent expansion of the meromorphic function f at p_j with respect to the coordinate t_j. The 2-cocycle ω in Proposition 1.1 satisfies

$$\sum_{j=1}^n \omega(X \otimes f_{p_j}, Y \otimes g_{p_j}) = 0$$

for any $X \otimes f, Y \otimes g \in \mathfrak{g}(p_1, \cdots, p_n)$ since

$$\sum_{j=1}^{n} \omega(X \otimes f_{p_j}, Y \otimes g_{p_j}) = \langle X, Y \rangle \sum_{j=1}^{n} \mathrm{Res}_{t_j=0}(df\ g)$$

and the sum of the residues of a meromorphic 1-form is zero. It follows that

$$\Delta([X,Y] \otimes fg) = [\Delta(X \otimes f), \Delta(Y \otimes g)]$$

holds. This completes the proof.   □

The *space of conformal blocks* $\mathcal{H}(p_1, \cdots, p_n; \lambda_1, \cdots, \lambda_n)$ is defined to be the space of linear forms

$$\Psi : H_{\lambda_1} \otimes \cdots \otimes H_{\lambda_n} \to \mathbf{C}$$

which are invariant under the diagonal action Δ of the Lie algebra $\mathfrak{g}(p_1, \cdots, p_n)$. In other words, $\mathcal{H}(p_1, \cdots, p_n; \lambda_1, \cdots, \lambda_n)$ consists of multilinear maps $\Psi : H_{\lambda_1} \times \cdots \times H_{\lambda_n} \to \mathbf{C}$ such that

$$\sum_{j=1}^{n} \Psi(\xi_1, \cdots, (X \otimes f_{p_j})\xi_j, \cdots, \xi_n) = 0$$

is satisfied for any $\xi_1 \in H_{\lambda_1}, \cdots, \xi_n \in H_{\lambda_n}$ and $X \otimes f \in \mathfrak{g}(p_1, \cdots, p_n)$. Here f_{p_j} is the Laurent expansion of f at p_j with respect to t_j. The space of conformal blocks $\mathcal{H}(p_1, \cdots, p_n; \lambda_1, \cdots, \lambda_n)$ is also expressed as

$$\mathrm{Hom}_{\mathfrak{g}(p_1, \cdots, p_n)}\left(\bigotimes_{j=1}^{n} H_{\lambda_j}, \mathbf{C}\right)$$

where $\mathbf{C}$ is considered to be a trivial module over $\mathfrak{g}(p_1, \cdots, p_n)$, i.e., any element of $\mathfrak{g}(p_1, \cdots, p_n)$ acts as zero on $\mathbf{C}$.

Let us consider the meromorphic function

$$f(z) = (z - z_i)^r, \quad r < 0,$$

defined on $\mathbf{C}P^1$. The function f has a pole only at p_j and it is also written as t_i^r using the coordinate function t_i. We express the Taylor expansion of f at p_j, $j \neq i$, as

$$f_{p_j}(t_j) = \sum_{m=0}^{\infty} a_m^{(j)} t_j^m.$$

Since an element Ψ of the space of conformal blocks

$$\mathcal{H}(p_1, \cdots, p_n; \lambda_1, \cdots, \lambda_n)$$

is invariant under the diagonal action of $X \otimes f$ for any $X \in \mathfrak{g}$ the equality

$$(1.19) \qquad \Psi(\xi_1, \cdots, (X \otimes t_i^r)\xi_i, \cdots, \xi_n)$$
$$= - \sum_{j, j \neq i} \sum_{m \geq 0} a_m^{(j)} \Psi(\xi_1, \cdots, (X \otimes t_j^m)\xi_j, \cdots, \xi_n)$$

holds for any $\xi_j \in H_j, 1 \leq j \leq n$. The above equality will often be useful in the following argument.

Let us now show that the space of conformal blocks is embedded in a space of coinvariant tensors of finite dimensional representations of $\mathfrak{g}$. The finite dimensional irreducible representation V_{λ_j} of $\mathfrak{g}$ with highest weight λ_j is regarded as a subspace of H_{λ_j}. For $\Psi \in \mathcal{H}(p_1, \cdots, p_n; \lambda_1, \cdots, \lambda_n)$, composing with the natural injective map

$$i : \bigotimes_{j=1}^{n} V_{\lambda_j} \to \bigotimes_{j=1}^{n} H_{\lambda_j}$$

we obtain a linear map

$$i^* \Psi : \bigotimes_{j=1}^{n} V_{\lambda_j} \to \mathbf{C}$$

defined by $i^* \Psi = \Psi \circ i$. This is the restriction of the multilinear form Ψ on $\bigotimes_{j=1}^{n} V_{\lambda_j}$ and we write Ψ_0 for $i^* \Psi$.

LEMMA 1.17. *For* $\Psi \in \mathcal{H}(p_1, \cdots, p_n; \lambda_1, \cdots, \lambda_n)$, *its restriction* Ψ_0 *on* $\bigotimes_{j=1}^{n} V_{\lambda_j}$ *is invariant under the diagonal action of* $\mathfrak{g}$. *Moreover, the linear map*

$$i^* : \mathcal{H}(p_1, \cdots, p_n; \lambda_1, \cdots, \lambda_n) \to \mathrm{Hom}_{\mathfrak{g}} \left(\bigotimes_{j=1}^{n} V_{\lambda_j}, \mathbf{C} \right)$$

thus obtained is injective. Namely, Ψ *is determined uniquely by* Ψ_0.

PROOF. Since Ψ is invariant under the diagonal action of

$$\mathfrak{g}(p_1, \cdots, p_n)$$

it is invariant, in particular, under the diagonal action of $X \otimes 1_{\mathbf{C}P^1}$ for any $X \in \mathfrak{g}$, where $1_{\mathbf{C}P^1}$ is a constant function on $\mathbf{C}P^1$ with value equal to 1. This implies

$$\sum_{j=1}^{n} \Psi(\xi_1 \otimes \cdots \otimes X\xi_j \otimes \cdots \otimes \xi_n) = 0$$

for any $X \in \mathfrak{g}$ and $\xi_j \in V_{\lambda_j}, 1 \leq j \leq n$. To show the injectivity of i^* we set

$$\mathcal{F}_d = \bigoplus_{d_1 + \cdots + d_n \leq d} \left(\bigotimes_{j=1}^{n} H_{\lambda_j}(d_j) \right)$$

using the direct sum decomposition $H_{\lambda_j} = \bigoplus_{d \geq 0} H_{\lambda_j}(d)$ introduced in (1.11). We have $\mathcal{F}_0 = \bigotimes_{j=1}^{n} V_{\lambda_j}$ and

$$\mathcal{F}_0 \subset \mathcal{F}_1 \subset \cdots \subset \mathcal{F}_d \subset \cdots$$

defines an increasing filtration on $\bigotimes_{j=1}^{n} H_{\lambda_j}$. Let us suppose that Ψ is identically zero on $\mathcal{F}_0$. We show by induction with respect to d that Ψ vanishes on $\mathcal{F}_d$ for any $d \geq 0$. It is enough to show that $\Psi(\xi) = 0$ holds for $\xi \in \mathcal{F}_d$ expressed as

$$\xi = \eta_1 \otimes \cdots \otimes (X \otimes t_i^r)\eta_i \otimes \cdots \otimes \eta_n,$$
$$X \in \mathfrak{g}, \quad r < 0, \quad \eta_1 \otimes \cdots \otimes \eta_n \in \mathcal{F}_{d-1}.$$

We see that this follows from the inductive hypothesis by means of the equation (1.19). This completes the proof. □

It follows, in particular, from the above lemma that the space of conformal blocks $\mathcal{H}(p_1, \cdots, p_n; \lambda_1, \cdots, \lambda_n)$ is a finite dimensional vector space. Our next object is to determine the image of the restriction map i^*. As explained in Section 1.2, the integrable highest weight module H_λ is the quotient of the Verma module M_λ by the submodule generated by the null vector $\chi = (E \otimes t^{-1})^d v$ where $d = k - \lambda + 1$ and $v \in V_\lambda$ is the highest weight vector. The existence of such a null vector imposes a further algebraic constraint on Ψ_0 for $\Psi \in \mathcal{H}(p_1, \cdots, p_n; \lambda_1, \cdots, \lambda_n)$ in addition to the condition that $\Psi_0 : V_{\lambda_1} \times \cdots \times V_{\lambda_n} \to \mathbf{C}$ must be invariant under the diagonal action of the Lie algebra $\mathfrak{g}$. More explicitly, we have the following proposition. We set $d_i = k - \lambda_i + 1, 1 \leq i \leq n$.

PROPOSITION 1.18. *If Ψ belongs to the above space of conformal blocks, then the multilinear map $\Psi_0 : V_{\lambda_1} \times \cdots \times V_{\lambda_n} \to \mathbf{C}$ obtained as a restriction of Ψ satisfies*

$$(1.20) \quad \Psi_0 \left(E^{m_1}\xi_1, \cdots, E^{m_{i-1}}\xi_{i-1}, v_i, E^{m_{i+1}}\xi_{i+1}, \cdots, E^{m_n}\xi_n \right) = 0$$

for each highest weight vector $v_i \in V_{\lambda_i}$, $\xi_j \in V_{\lambda_j}$, $j \neq i, 1 \leq j \leq n$, and any non-negative integers $m_j, j \neq i, 1 \leq j \leq n$, with $\sum_{j, j \neq i} m_j = d_i$.

PROOF. It is enough to show the statement in the case $i = 1$. Since $\chi_1 = (E \otimes t_1^{-1})^{d_1} v_1$ is a null vector we have

$$\Psi((E \otimes t_1^{-1})^{d_1} v_1, \xi_2, \cdots, \xi_n) = 0.$$

Applying the equation (1.19) to the left hand side of the above equality and restricting Ψ on $V_{\lambda_1} \times \cdots \times V_{\lambda_n}$, we obtain

$$\sum_{m_2 + \cdots + m_n = d_1} \frac{d_1!}{m_2! \cdots m_n!} \prod_{2 \leq j \leq n} (z_j - z_1)^{-m_j} f_{m_2, \cdots, m_n} = 0$$

where

$$f_{m_2, \cdots, m_n} = \Psi_0 (v_1, E^{m_2} \xi_2, \cdots, E^{m_n} \xi_n).$$

The above equality holds for any $z_1, \cdots, z_n$, therefore, the coefficient of $\prod_{2 \leq j \leq n} (z_j - z_1)^{-m_j}$ must vanish. This shows the desired equality. □

The argument so far might be extended in a similar way to any complex semisimple Lie algebra $\mathfrak{g}$. In the following, we investigate the above algebraic constraint in more detail in the case $\mathfrak{g} = sl_2(\mathbf{C})$. We consider the case $n = 3$. We denote by $N_{\lambda_1 \lambda_2 \lambda_3}$ the dimension of the space of conformal blocks

$$\mathcal{H}(p_1, p_2, p_3; \lambda_1, \lambda_2, \lambda_3)$$

as a complex vector space.

PROPOSITION 1.19. *In the case $n = 3$, if the level k highest weights $\lambda_1, \lambda_2, \lambda_3 \in \mathbf{Z}$ satisfy the conditions*

$$\lambda_1 + \lambda_2 + \lambda_3 \in 2\mathbf{Z},$$
$$|\lambda_1 - \lambda_2| \leq \lambda_3 \leq \lambda_1 + \lambda_2,$$
$$\lambda_1 + \lambda_2 + \lambda_3 \leq 2k,$$

then the dimension $N_{\lambda_1 \lambda_2 \lambda_3}$ of the space of conformal blocks

$$\mathcal{H}(p_1, p_2, p_3; \lambda_1, \lambda_2, \lambda_3)$$

is equal to 1. Otherwise, $N_{\lambda_1 \lambda_2 \lambda_3} = 0$.

PROOF. As explained in Proposition 1.4, it is known that

$$\mathrm{Hom}_{sl_2(\mathbf{C})}(V_{\lambda_1} \otimes V_{\lambda_2} \otimes V_{\lambda_3}, \mathbf{C}) \cong \mathbf{C}$$

holds if and only if the condition

$$\lambda_1 + \lambda_2 + \lambda_3 \in 2\mathbf{Z},$$
$$|\lambda_1 - \lambda_2| \leq \lambda_3 \leq \lambda_1 + \lambda_2$$

is satisfied. This is called the Clebsch-Gordan condition. If the Clebsch-Gordan condition is not satisfied, we have

$$\mathrm{Hom}_{sl_2(\mathbf{C})}(V_{\lambda_1} \otimes V_{\lambda_2} \otimes V_{\lambda_3}, \mathbf{C}) = 0.$$

We have shown in Lemma 1.17 that the space of conformal blocks $\mathcal{H}(p_1, p_2, p_3; \lambda_1, \lambda_2, \lambda_3)$ is embedded as a subspace of

$$\mathrm{Hom}_{sl_2(\mathbf{C})}(V_{\lambda_1} \otimes V_{\lambda_2} \otimes V_{\lambda_3}, \mathbf{C}).$$

Hence its dimension $N_{\lambda_1 \lambda_2 \lambda_3}$ is either 0 or 1. We suppose that the Clebsch-Gordan condition is satisfied and show that $N_{\lambda_1 \lambda_2 \lambda_3}$ is equal to 1 if and only if the constraint

$$\lambda_1 + \lambda_2 + \lambda_3 \leq 2k$$

coming from the level k is satisfied in addition to the Clebsch-Gordan condition. As in Section 1.2 we take H, E and F as a basis of $sl_2(\mathbf{C})$. Since the condition is symmetric in λ_1, λ_2 and λ_3, we apply Proposition 1.18 in the case $i = 1$. If Ψ belongs to $\mathcal{H}(p_1, p_2, p_3; \lambda_1, \lambda_2, \lambda_3)$, then we have

$$(1.21) \qquad \Psi_0(v_1, E^{m_2}\xi_2, E^{m_3}\xi_3) = 0, \quad m_2 + m_3 = k - \lambda_1 - 1,$$

for the highest weight vector $v_1 \in V_{\lambda_1}$ and any element $\xi_j \in V_{\lambda_j}, j = 2, 3$. We may suppose that ξ_2 and ξ_3 are eigenvectors of H. We denote by α_2 and α_3 the corresponding eigenvalues. The eigenvalues α_j, $j = 2, 3$, are integers with $-\lambda_j \leq \alpha_j \leq \lambda_j$. We write $\alpha_j = -\lambda_j + 2n_j$ with a non-negative integer n_j. We have

$$(1.22)$$
$$H(v_1 \otimes E^{m_2}\xi_2 \otimes E^{m_3}\xi_3)$$
$$= (2(k+1) - (\lambda_1 + \lambda_2 + \lambda_3) + 2(n_2 + n_3)) v_1 \otimes E^{m_2}\xi_2 \otimes E^{m_3}\xi_3.$$

If $\lambda_1 + \lambda_2 + \lambda_3 \leq 2k$, then we have $2(k+1) - (\lambda_1 + \lambda_2 + \lambda_3) + 2(n_2 + n_3) \neq 0$ and the condition (1.21) follows from the fact that Ψ_0 is invariant under the diagonal action of H shown in Lemma 1.17 together with the equality (1.22). Therefore the algebraic constraint in Proposition 1.18 imposes no condition on the space of coinvariant tensors $\mathrm{Hom}_{\mathfrak{g}}(V_{\lambda_1} \otimes V_{\lambda_2} \otimes V_{\lambda_3}, \mathbf{C})$. In the case $\lambda_1 + \lambda_2 + \lambda_3 > 2k$ the algebraic constraint in Proposition 1.18 is non-trivial. To show that

$$\Psi_0(\eta_1, \eta_2, \eta_3) = 0$$

holds for any $\eta_j \in V_{\lambda_j}$ if $\lambda_1 + \lambda_2 + \lambda_3 > 2k$, we suppose that $\eta_1 = v_1$ and that η_2 and η_3 are eigenvectors of H. Furthermore, we may assume that $H(\eta_1 \otimes \eta_2 \otimes \eta_3) = 0$ since if $H(\eta_1 \otimes \eta_2 \otimes \eta_3) \neq 0$, then we have $H(\eta_1 \otimes \eta_2 \otimes \eta_3) = \alpha\eta_1 \otimes \eta_2 \otimes \eta_3$ with $\alpha \neq 0$ and the invariance under H implies $\Psi_0(\eta_1, \eta_2, \eta_3) = 0$. We set $H\eta_2 = -\lambda_2\eta_2$ and $H\eta_3 = (\lambda_2 - \lambda_3)\eta_3$. It follows from $\lambda_1 + \lambda_2 + \lambda_3 > 2k$ that we have

$$\lambda_2 - \lambda_1 - 2d_1 = \lambda_1 + \lambda_2 - 2(k+1) \geq -\lambda_3.$$

This implies that η_3 can be written as $E^{d_1}\xi$ with some non-zero vector $\xi \in V_{\lambda_3}$. Therefore, we obtain $\Psi_0(v_1, \eta_2, \eta_3) = 0$ by the algebraic constraint (1.21). By means of this property and the $\mathfrak{g}$ invariance of Ψ_0 we can show inductively that Ψ_0 vanishes on $V_{\lambda_1} \otimes V_{\lambda_2} \otimes V_{\lambda_3}$. Hence we have $N_{\lambda_1\lambda_2\lambda_3} = 0$. This completes the proof. $\qquad\square$

The condition for λ_1, λ_2 and λ_3 so that $N_{\lambda_1\lambda_2\lambda_3} \neq 0$ in Proposition 1.19 is called the *quantum Clebsch-Gordan condition* at level k. In general, the method of counting dimensions of the space of conformal blocks from the given data of highest weights associated with points is called the *fusion rule*.

So far, we have assumed that none of the p_j, $1 \leq j \leq n$, coincides with ∞. In the next chapter, we often use the following type of space of conformal blocks. We take $n+1$ distinct points $p_1, \cdots, p_n, p_{n+1}$ so that $p_{n+1} = \infty$ and associate integrable highest weight modules $H_{\lambda_1}, \cdots, H_{\lambda_n}, H^*_{\lambda_{n+1}}$ of level k to the points $p_1, \cdots, p_n, p_{n+1}$ where $H^*_{\lambda_{n+1}}$ is the dual module of $H_{\lambda_{n+1}}$ in the sense of Section 1.2. We define the space of conformal blocks

$$\mathcal{H}(p_1, \cdots, p_n, p_{n+1}; \lambda_1, \cdots, \lambda_n, \lambda^*_{n+1})$$

by the space of coinvariant tensors

$$(1.23) \qquad \mathrm{Hom}_{\mathfrak{g}(p_1, \cdots, p_n, p_{n+1})}\left((\bigotimes_{j=1}^{n} H_{\lambda_j}) \otimes H^*_{\lambda_{n+1}}, \mathbf{C} \right).$$

Here the diagonal action of $\mathfrak{g}(p_1, \cdots, p_n, p_{n+1})$ is defined by the Laurent expansion at each point p_j, $1 \leq j \leq n+1$, as in the last paragraphs. Applying the argument of the proof of Lemma 1.17, we can show that this space of conformal blocks is embedded in

$$\mathrm{Hom}_{\mathfrak{g}}\left((\bigotimes_{j=1}^{n} V_{\lambda_j}) \otimes V^*_{\lambda_{n+1}}, \mathbf{C} \right).$$

The dimension of our space of conformal blocks as a complex vector space is denoted by $N^{\lambda_{n+1}}_{\lambda_1 \cdots \lambda_n}$.

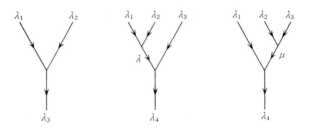

FIGURE 1.2. Basis of the space of conformal blocks represented by labelled trees

In the case $n = 1$,

$$\mathrm{Hom}_{\mathfrak{g}}(V_{\lambda_1} \otimes V_{\lambda_2}^*, \mathbf{C})$$

is isomorphic to $\mathbf{C}$ if $\lambda_1 = \lambda_2$ and is zero otherwise. The algebraic constraint (1.20) imposes no more conditions and we have $N_{\lambda_1}^{\lambda_2} = 1$ if $\lambda_1 = \lambda_2$ and $N_{\lambda_1}^{\lambda_2} = 0$ if $\lambda_1 \neq \lambda_2$. In the case $n = 2$, $N_{\lambda_1 \lambda_2}^{\lambda_3} = 1$ if λ_1, λ_2 and λ_3 satisfy the quantum Clebsch-Gordan condition at level k and $N_{\lambda_1 \lambda_2}^{\lambda_3} = 0$ otherwise. If we formulate the space of conformal blocks for a general complex semisimple Lie algebra, then we have $N_{\lambda_1 \lambda_2}^{\lambda_3} = N_{\lambda_1 \lambda_2 \lambda_3 *}$ where we denote by λ^* the highest weight of the dual of the representation with highest weight λ. In the case $\mathfrak{g} = sl_2(\mathbf{C})$, any finite dimensional irreducible representation of $\mathfrak{g}$ is equivalent to its dual representation and we have $N_{\lambda_1 \lambda_2}^{\lambda_3} = N_{\lambda_1 \lambda_2 \lambda_3}$. In the case $n = 2$, the space of conformal blocks (1.23) is embedded in

$$\mathrm{Hom}_{\mathfrak{g}}(V_{\lambda_1} \otimes V_{\lambda_2}, V_{\lambda_3}).$$

We suppose that λ_1, λ_2 and λ_3 satisfy the quantum Clebsch-Gordan condition at level k. We have a basis of the above vector space corresponding to a projection map $p_{\lambda_1 \lambda_2}^{\lambda_3} : V_{\lambda_1} \otimes V_{\lambda_2} \to V_{\lambda_3}$ with respect to the decomposition of the tensor representation $V_{\lambda_1} \otimes V_{\lambda_2}$ into irreducible components. This determines a basis of the space of conformal blocks in the case $n = 2$ up to a constant multiple. We represent this basis by the graph depicted in Figure 1.2.

In general, a basis of the space of conformal blocks can be described in terms of *labelled trees*. Here, by a tree we mean a contractible 1-dimensional finite simplicial complex embedded in a plane. The valency of a vertex p of a tree is the number of edges meeting at p. We suppose that a tree Γ_n satisfies the following properties.

1. Each vertex of Γ_n has valency 3 or 1.
2. The number of vertices of Γ_n with valency 1 is equal to $n + 1$.

An edge of Γ_n is called external if the edge is adjacent to a vertex with valency 1. We have $n + 1$ external edges. We fix an ordering of the external edges and call them $\gamma_1, \cdots, \gamma_{n+1}$. We denote by $P_+(k)$ the set of level k highest weights $\{0, 1, \cdots, k\}$. We fix $\lambda_1, \cdots, \lambda_{n+1} \in P_+(k)$. A labelling of the tree Γ_n is a map f from the set of edges of Γ_{n+1} to $P_+(k)$. We suppose $f(\gamma_j) = \lambda_j, 1 \leq j \leq n + 1$. We say that f is *admissible* if for each vertex p with valency 3, the labels $f(e_{i_1}), f(e_{i_2}), f(e_{i_3})$ for the edges $e_{i_1}, e_{i_2}, e_{i_3}$ meeting at p satisfy the quantum Clebsch-Gordan condition at level k.

In the case $n = 3$, the space of conformal blocks (1.23) embedded in

$$\mathrm{Hom}_{\mathfrak{g}}(V_{\lambda_1} \otimes V_{\lambda_2} \otimes V_{\lambda_3}, V_{\lambda_4})$$

has a basis described by a tree with admissible labellings in the following way. Consider the composition of projection maps

$$p_{\lambda_1 \lambda_2}^{\lambda} \otimes id_{V_{\lambda_3}} : (V_{\lambda_1} \otimes V_{\lambda_2}) \otimes V_{\lambda_3} \to V_\lambda \otimes V_{\lambda_3},$$

$$p_{\lambda \lambda_3}^{\lambda_4} : V_\lambda \otimes V_{\lambda_3} \to V_{\lambda_4}.$$

The $\mathfrak{g}$ homomorphism obtained for each admissible labelling λ forms a basis of our space of conformal blocks. This basis is represented as the tree with label λ in Figure 1.2. The tree with label μ in Figure 1.2 corresponds to another basis. In general, using an argument similar to that in the proof of Proposition 1.19 based on the algebraic constraint in Proposition 1.18, we have the following proposition.

PROPOSITION 1.20. *The space of conformal blocks (1.23) has a basis which is in one-to-one correspondence with the set of admissible labellings f for Γ_n such that $f(\gamma_j) = \lambda_j, 1 \leq j \leq n + 1$.*

REMARK 1.21. In the above argument, we fixed a positive integer k as a level and treated integrable highest weight modules of level k. If we consider the case when k is a generic parameter, then the Verma module $M_{k,\lambda}$ is irreducible and we can define the space of conformal blocks based on Verma modules. Since the algebraic constraint in Proposition 1.18 is derived from the existence of a null vector, there is no such constraint in the case of Verma modules and the condition for $N_{\lambda_1 \lambda_2 \lambda_3} \neq 0$ coincides with the classical Clebsch-Gordan condition.

1.5. KZ equation

In this section, we proceed to discuss a family of the space of conformal blocks as the points $p_1, \cdots, p_n$ move on $\mathbf{C}P^1$. Our main object is to show that this family forms a vector bundle with a natural flat connection. First, we define the *configuration space* of n ordered distinct points on $\mathbf{C}P^1$ by

$$\mathrm{Conf}_n(\mathbf{C}P^1) = \{(p_1, \cdots, p_n) \in (\mathbf{C}P^1)^n \mid p_i \neq p_j \text{ if } i \neq j\}$$

where $(\mathbf{C}P^1)^n$ denotes the n-fold Cartesian product of $\mathbf{C}P^1$. We fix a positive integer k called the level as in the last section and choose level k highest weights $\lambda_1, \cdots, \lambda_n$. We will consider the disjoint union of the space of conformal blocks

$$\mathcal{E}_{\lambda_1, \cdots, \lambda_n} = \bigcup_{(p_1, \cdots, p_n) \in \mathrm{Conf}_n(\mathbf{C}P^1)} \mathcal{H}(p_1, \cdots, p_n; \lambda_1, \cdots, \lambda_n)$$

as a vector bundle over $\mathrm{Conf}_n(\mathbf{C}P^1)$.

We consider the projection

$$\pi : (\mathbf{C}P^1)^{n+1} \to (\mathbf{C}P^1)^n$$

on the first n factors. For $j = 1, \cdots, n$ we consider the hyperplane D_j of $(\mathbf{C}P^1)^{n+1}$ defined by $z_{n+1} = z_j$ where $(z_1, \cdots, z_{n+1})$ is a coordinate function for $(\mathbf{C}P^1)^{n+1}$. Here we fix a global coordinate function on $\mathbf{C}P^1$ as in Section 1.3. For n distinct points $p_1, \cdots, p_n$ on $\mathbf{C}P^1$ with coordinates $z_1, \cdots, z_n$ the intersection of $\pi^{-1}(p_1, \cdots, p_n) \cong \mathbf{C}P^1$ and $D_j, 1 \leq j \leq n$, has a coordinate z_j with respect to the coordinate function z_{n+1}. Thus we have a family of $\mathbf{C}P^1$ with n distinct points over the configuration space $\mathrm{Conf}_n(\mathbf{C}P^1)$.

In order to put the structure of a vector bundle on $\mathcal{E}_{\lambda_1 \cdots \lambda_n}$ we start with a trivial vector bundle

$$E = \mathrm{Conf}_n(\mathbf{C}P^1) \times \mathrm{Hom}_{\mathbf{C}}\left(\bigotimes_{j=1}^n H_{\lambda_j}, \mathbf{C}\right)$$

over $\mathrm{Conf}_n(\mathbf{C}P^1)$ with fibre $\mathrm{Hom}_{\mathbf{C}}(\bigotimes_{j=1}^n H_{\lambda_j}, \mathbf{C})$ and consider $\mathcal{E}_{\lambda_1 \cdots \lambda_n}$ as a subset of E. Let U be an open set of $\mathrm{Conf}_n(\mathbf{C}P^1)$ and denote by $\mathcal{M}_{D_1 \cdots D_n}(U)$ the set of meromorphic functions on $\pi^{-1}(U)$ with poles of any order at most along $D_1, \cdots, D_n$. We see that

$$\mathfrak{g} \otimes \mathcal{M}_{D_1 \cdots D_n}(U)$$

has the structure of a Lie algebra in a natural way. A meromorphic function $f \in \mathfrak{g} \otimes \mathcal{M}_{D_1 \cdots D_n}(U)$ has a Laurent expansion along D_j,

$1 \leq j \leq n$, of the form

$$f_{D_j}(t_j) = \sum_{m=-N}^{\infty} a_m(z_1, \cdots, z_n) t_j^m$$

where $t_j = z_{n+1} - z_j$ and $a_m(z_1, \cdots, z_n)$ is a holomorphic function in $z_1, \cdots, z_n$ with values in $\mathfrak{g}$. We denote by $\mathcal{O}(U)$ the set of holomorphic functions on U. The above Laurent expansion gives a map

$$\tau_j : \mathfrak{g} \otimes \mathcal{M}_{D_1 \cdots D_n}(U) \to \mathfrak{g} \otimes \mathcal{O}(U) \otimes \mathbf{C}((t_j))$$

determined by $\tau_j(f) = f_{D_j}(t_j)$. For a fixed $(p_1, \cdots, p_n) \in U$ with coordinates $(z_1, \cdots, z_n)$ we regard the Laurent series $f_{D_j}(t_j)$ as an element of the loop algebra $\mathfrak{g} \otimes \mathbf{C}((t_j))$ and we consider its action on the integrable highest weight module H_{λ_j}. This construction permits us to define the diagonal action of the Lie algebra $\mathfrak{g} \otimes \mathcal{M}_{D_1 \cdots D_n}(U)$ on the tensor product $H_{\lambda_1} \otimes \cdots \otimes H_{\lambda_n}$. Let us notice that this diagonal action depends on $(p_1, \cdots, p_n) \in U$.

We define $\mathcal{E}_{\lambda_1 \cdots \lambda_n}(U)$ to be the set of smooth sections

$$\Psi : U \to E$$

of the trivial vector bundle E such that $\Psi(p_1, \cdots, p_n)$ is invariant under the above diagonal action of $\mathfrak{g} \otimes \mathcal{M}_{D_1 \cdots D_n}(U)$ defined at any $(p_1, \cdots, p_n) \in U$. Namely, $\Psi \in \mathcal{E}_{\lambda_1 \cdots \lambda_n}(U)$ satisfies

$$\sum_{j=1}^{n} [\Psi(p_1, \cdots, p_n)] (\xi_1, \cdots, \tau_j(f)\xi_j, \cdots, \xi_n) = 0$$

for any $f \in \mathfrak{g} \otimes \mathcal{M}_{D_1 \cdots D_n}(U)$ and $\xi_j \in H_{\lambda_j}$, $1 \leq j \leq n$, at any $(p_1, \cdots, p_n) \in U$. If we fix $(p_1, \cdots, p_n) \in U$, then $\Psi(p_1, \cdots, p_n)$ is an element of the space of conformal blocks $\mathcal{H}(p_1, \cdots, p_n; \lambda_1, \cdots, \lambda_n)$. This construction for any open set $U \subset \mathrm{Conf}_n(\mathbf{CP}^1)$ gives the sheaf of smooth sections of $\mathcal{E}_{\lambda_1 \cdots \lambda_n}$.

Let us now show that $\mathcal{E}_{\lambda_1 \cdots \lambda_n}$ has the structure of a vector bundle with a flat connection on the open subset

$$\mathrm{Conf}_n(\mathbf{C}) = \{(z_1, \cdots, z_n) \in \mathbf{C}^n \mid z_i \neq z_j \text{ if } i \neq j\}$$

of $\mathrm{Conf}_n(\mathbf{CP}^1)$ by means of the Sugawara operator L_{-1} introduced in the equation (1.9) of Section 1.2. Let $\Psi : H_{\lambda_1} \otimes \cdots \otimes H_{\lambda_n} \to \mathbf{C}$ be a multilinear map. For an element X of the affine Lie algebra $\widehat{\mathfrak{g}}$ or its universal enveloping algebra we define a multilinear map $X^{(j)}\Psi : H_{\lambda_1} \otimes \cdots \otimes H_{\lambda_n} \to \mathbf{C}$ by

$$[X^{(j)}\Psi](\xi_1, \cdots, \xi_n) = \Psi(\xi_1, \cdots, X\xi_j, \cdots, \xi_n),$$

where $\xi_1 \in H_{\lambda_1}, \cdots, \xi_n \in H_{\lambda_n}$ and X acts on the j-th component. With this notation we have the following proposition.

PROPOSITION 1.22. *If Ψ is a smooth section of $\mathcal{E}_{\lambda_1 \cdots \lambda_n}$ over an open set $U \subset \mathrm{Conf}_n(\mathbf{C})$, then*

$$\frac{\partial \Psi}{\partial z_j} - L_{-1}^{(j)} \Psi$$

is a smooth section of $\mathcal{E}_{\lambda_1 \cdots \lambda_n}$ over U for each $j, 1 \leq j \leq n$, as well.

PROOF. For $f \in \mathfrak{g} \otimes \mathcal{M}_{D_1 \cdots D_n}(U)$ we express its Laurent expansion along D_j as

$$\tau_j(f) = f_{D_j}(t_j) = \sum_{m=-N}^{\infty} a_m(z_1, \cdots, z_n) t_j^m$$

where $t_j = z_{n+1} - z_j$. Its partial derivative $f_{z_j} = \frac{\partial f}{\partial z_j}$ also belongs to $\mathfrak{g} \otimes \mathcal{M}_{D_1 \cdots D_n}(U)$. The Laurent expansion of f_{z_j} along D_j is given by

$$\tau_j(f_{z_j}) = \sum_{m=-N}^{\infty} \left(\frac{\partial a_m}{\partial z_j} t_j^m - a_m m\, t_j^{m-1} \right).$$

We introduce the operator

$$\partial_j : \mathfrak{g} \otimes \mathcal{O}(U) \to \mathfrak{g} \otimes \mathcal{O}(U), \ 1 \leq j \leq n,$$

defined by $\partial_j(h) = h_{z_j}$ and we extend ∂_j to a linear operator on $\mathfrak{g} \otimes \mathcal{O}(U) \otimes \mathbf{C}((t_j))$ so that ∂_j acts trivially on $\mathbf{C}((t_j))$. With this notation we have

$$\tau_j(f_{z_j}) = \partial_j \tau_j(f) - \frac{\partial}{\partial t_j} \tau_j(f).$$

Moreover, since we have

$$\frac{\partial}{\partial t_j} \tau_j(f) = -[L_{-1}, \tau_j(f)]$$

by Proposition 1.6, we obtain

(1.24) $$\tau_j(f_{z_j}) = \partial_j \tau_j(f) + [L_{-1}, \tau_j(f)].$$

Let us notice that we have $\tau_i(f_{z_j}) = \partial_j \tau_i(f)$ for $i \neq j$.

We have to show that

$$\sum_{i=1}^{n} \left(\frac{\partial \Psi}{\partial z_j} - L_{-1}^{(j)} \Psi \right) (\xi_1, \cdots, \tau_i(f)\xi_i, \cdots, \xi_n) = 0$$

holds for any $f \in \mathfrak{g} \otimes \mathcal{M}_{D_1 \cdots D_n}(U)$ and $\xi_1 \in H_{\lambda_1}, \cdots, \xi_n \in H_{\lambda_n}$. We have

$$\frac{\partial}{\partial z_j} \left[\Psi \left(\xi_1, \cdots, \tau_i(f)\xi_i, \cdots, \xi_n \right) \right]$$

$$= \left(\frac{\partial \Psi}{\partial z_j} \right) (\xi_1, \cdots, \tau_i(f)\xi_i, \cdots, \xi_n) + \Psi \left(\xi_1, \cdots, \partial_j \tau_i(f)\xi_i, \cdots, \xi_n \right).$$

Combining the above equality, the equation (1.24) and the definition of Ψ we have

$$\sum_{i=1}^{n} \left(\frac{\partial \Psi}{\partial z_j} - L_{-1}^{(j)} \Psi \right) (\xi_1, \cdots, \tau_i(f)\xi_i, \cdots, \xi_n)$$

$$= \sum_{i=1}^{n} \left(\frac{\partial}{\partial z_j} \left[\Psi \left(\xi_1, \cdots, \tau_i(f)\xi_i, \cdots, \xi_n \right) \right] \right.$$

$$\left. - \Psi \left(\xi_1, \cdots, \tau_i(f_{z_j})\xi_i, \cdots, \xi_n \right) \right).$$

Since f_{z_j} belongs to $\mathfrak{g} \otimes \mathcal{M}_{D_1 \cdots D_n}(U)$ we conclude that the right hand side of the above equality vanishes by the definition of Ψ. This completes the proof. □

The above proposition leads us to introduce a linear operator

$$\nabla_{\frac{\partial}{\partial z_j}} : \mathcal{E}_{\lambda_1 \cdots \lambda_n}(U) \to \mathcal{E}_{\lambda_1 \cdots \lambda_n}(U)$$

defined by

$$\nabla_{\frac{\partial}{\partial z_j}} \Psi = \frac{\partial \Psi}{\partial z_j} - L_{-1}^{(j)} \Psi.$$

THEOREM 1.23. *The family of conformal blocks $\mathcal{E}_{\lambda_1 \cdots \lambda_n}$ over the configuration space $\mathrm{Conf}_n(\mathbf{C})$ has the structure of a vector bundle with a flat connection.*

PROOF. We set $M = \mathrm{Conf}_n(\mathbf{C})$ and we first consider a trivial vector bundle $E = M \times F$ over M where the fibre F is $\mathrm{Hom}_{\mathbf{C}}(\bigotimes_{j=1}^{n} H_{\lambda_j}, \mathbf{C})$. The $\mathbf{C}$ linear map

$$\nabla : \Gamma(E) \to \Gamma(T^* M_{\mathbf{C}} \otimes E)$$

given by

(1.25) $$\nabla \Psi = d\Psi - \sum_{i=1}^{n} L_{-1}^{(i)} \Psi dz_i$$

is a connection on E. We put

$$\omega = \sum_{i=1}^{n} L_{-1}^{(i)} dz_i,$$

which is considered to be a 1-form on M with values in $\mathrm{End}(F)$. Since $L_{-1}^{(i)}$ does not depend on $z \in M$ we have $d\omega = 0$. Moreover, we have $\omega \wedge \omega = 0$ since

$$[L_{-1}^{(i)}, L_{-1}^{(j)}] = 0, \quad 1 \le i, j \le n,$$

holds. It follows that the curvature of the connection ∇ vanishes. In other words ∇ is a flat connection on E. Proposition 1.22 shows that the connection ∇ restricts to the subspace $\mathcal{E}_{\lambda_1 \cdots \lambda_n} \subset E$. We have local frames of $\mathcal{E}_{\lambda_1 \cdots \lambda_n}$ as horizontal sections with respect to the connection ∇. This gives a local trivialization of $\mathcal{E}_{\lambda_1 \cdots \lambda_n}$, and consequently we have a structure of a vector bundle on $\mathcal{E}_{\lambda_1 \cdots \lambda_n}$. The connection ∇ restricted to $\mathcal{E}_{\lambda_1 \cdots \lambda_n}$ is flat. This completes the proof. $\square$

We call $\mathcal{E}_{\lambda_1 \cdots \lambda_n}$ the *conformal block bundle*. For

$$\Psi_0 : V_{\lambda_1} \otimes \cdots \otimes V_{\lambda_n} \to \mathbf{C}$$

obtained as the restriction of a multilinear form $\Psi : H_{\lambda_1} \otimes \cdots \otimes H_{\lambda_n} \to \mathbf{C}$ contained in the space of conformal blocks, we give a more explicit description of the above connection ∇. We denote by $\{I_\mu\}$ an orthonormal basis of the Lie algebra $\mathfrak{g}$ with respect to the Cartan-Killing form and we set

$$\Omega = \sum_{\mu} I_\mu \otimes I_\mu.$$

For i, j, $1 \le i \ne j \le n$, we denote by $\Omega^{(ij)} \Psi_0$ the action of Ω on Ψ on the i-th and j-th components. Namely, for $\xi_j \in V_{\lambda_j}, 1 \le j \le n$, $\Omega^{(ij)} \Psi_0$ is given by

$$[\Omega^{(ij)} \Psi_0](\xi_1, \cdots, \xi_n)$$
$$= \sum_{\mu} \Psi_0(\xi_1, \cdots, I_\mu \xi_i, \cdots, I_\mu \xi_j, \cdots, \xi_n).$$

It follows from the equation (1.19) that we have

$$(1.26) \qquad [(X \otimes t^{-1})^{(i)} \Psi](\xi_1, \cdots, \xi_n)$$
$$= \sum_{j, j \ne i} (z_i - z_j)^{-1} \Psi(\xi_1, \cdots, X\xi_j, \cdots, \xi_n).$$

PROPOSITION 1.24. *If a multilinear form* $\Psi : H_{\lambda_1} \otimes \cdots \otimes H_{\lambda_n} \to$
C *belongs to the space of conformal blocks* $\mathcal{H}(p_1, \cdots, p_n; \lambda_1, \cdots, \lambda_n)$,
then the restriction $\left(L_{-1}^{(i)} \Psi \right)_0$ *of* $L_{-1}^{(i)} \Psi : H_{\lambda_1} \otimes \cdots \otimes H_{\lambda_n} \to$ **C** *on*
$V_{\lambda_1} \otimes \cdots \otimes V_{\lambda_n}$ *is expressed as*

$$\left(L_{-1}^{(i)} \Psi \right)_0 = \sum_{j, j \neq i} \frac{\Omega^{(ij)} \Psi_0}{z_i - z_j}.$$

The action of $L_n^{(i)}, n \geq 0$, *satisfies*

$$\left(L_n^{(i)} \Psi \right)_0 = 0, \quad n > 0,$$
$$\left(L_0^{(i)} \Psi \right)_0 = \Delta_{\lambda_i} \Psi_0$$

where Δ_{λ_i} *is the eigenvalue of* L_0 *on the irreducible representation*
V_{λ_i}.

PROOF. By the definition of the Sugawara operator given in
equation (1.9) of Section 1.2 we have

$$L_{-1} v = \frac{1}{k+2} \left(\sum_{\mu} I_\mu \otimes t^{-1} \cdot I_\mu \right) v$$

for $v \in V_{\lambda_j} \subset H_{\lambda_j}$. Combining with the equality (1.26), we obtain

$$\sum_{\mu} \Psi \left(\xi_1, \cdots, (I_\mu \otimes t^{-1} \cdot I_\mu) \xi_i, \cdots, \xi_n \right)$$
$$= \sum_{j, j \neq i} (z_i - z_j)^{-1} \Psi_0 \left(\xi_1, \cdots, I_\mu \xi_i, \cdots, I_\mu \xi_j, \cdots, \xi_n \right),$$

which shows the desired expression for L_{-1}. It follows directly from
the definition of the Sugawara operators that we have

$$L_n \xi_i = 0, \quad n > 0,$$
$$L_0 \xi_i = \Delta_{\lambda_i} \xi_i$$

for $\xi_i \in V_{\lambda_i}$. The statement for L_n, $n \geq 0$, follows immediately. $\square$

Combining Theorem 1.23 and Proposition 1.24, we obtain the
following result.

THEOREM 1.25. *Let* Ψ *be a horizontal section of the conformal
block bundle* $\mathcal{E}_{\lambda_1 \cdots \lambda_n}$. *Then the restriction* Ψ_0 *of* Ψ *on* $V_{\lambda_1} \otimes \cdots \otimes V_{\lambda_n}$

satisfies the system of partial differential equations

$$\frac{\partial \Psi_0}{\partial z_i} = \frac{1}{k+2} \sum_{j, j \neq i} \frac{\Omega^{(ij)} \Psi_0}{z_i - z_j}, \quad 1 \leq i \leq n.$$

The above differential equation for Ψ_0 is called the *Knizhnik-Zamolodchikov equation*, which will often be abbreviated to the *KZ equation*. The connection ∇ for the conformal block bundle is called the *KZ connection*.

REMARK 1.26. In the case when we consider an extra point $p_{n+1} = \infty$ with highest weight λ_{n+1}^*, one can define the conformal block bundle over $\mathrm{Conf}_n(\mathbf{C})$ in a similar way as above. The KZ equation has the same form as in Theorem 1.25 where Ψ_0 is considered to be an element of

$$\mathrm{Hom}_{\mathfrak{g}} \left(\bigotimes_{j=1}^{n} V_{\lambda_j}, V_{\lambda_{n+1}} \right).$$

We now show that the horizontal sections of the conformal block bundle with respect to the KZ connection satisfy a certain invariance under the Möbius transformations of the Riemann sphere. The following proposition describes the invariance at the infinitesimal level.

PROPOSITION 1.27. *Let Ψ be a horizontal section of the conformal block bundle $\mathcal{E}_{\lambda_1 \cdots \lambda_n}$. Then its restriction Ψ_0 on $V_{\lambda_1} \otimes \cdots \otimes V_{\lambda_n}$ satisfies*

$$\sum_{i=1}^{n} z_i^r \left(z_i \frac{\partial}{\partial z_i} + (r+1) \Delta_{\lambda_i} \right) \Psi_0 = 0$$

for $r = -1, 0, 1$.

PROOF. Since Ψ_0 is invariant under the diagonal action of $\mathfrak{g}$ we have

$$\sum_{j=1}^{n} \Omega^{(ij)} \Psi_0 = 0, \quad 1 \leq i \leq n.$$

Taking the sum with respect to i, $1 \leq i \leq n$, of the above equality we have

$$\sum_{i \neq j} \Omega^{(ij)} \Psi_0 = - \sum_{j=1}^{n} \Omega^{(jj)} \Psi_0,$$

from which we obtain

$$(1.27) \qquad \sum_{1 \le i < j \le n} \Omega^{(ij)} \Psi_0 = -(k+2) \sum_{j=1}^{n} \Delta_{\lambda_j} \Psi_0$$

by using the fact that $\Omega^{(jj)}$ is the Casimir operator and the symmetry $\Omega^{(ij)} = \Omega^{(ji)}$. Since Ψ is a horizontal section we have

$$(1.28) \qquad \sum_{i=1}^{n} z_i^r \frac{\partial \Psi_0}{\partial z_i} = \frac{1}{k+2} \sum_{i \neq j} \frac{z_i^r \Omega^{(ij)} \Psi_0}{z_i - z_j}.$$

In the case $r = -1$ the right hand side of the equality (1.28) is zero by the symmetry $\Omega^{(ij)} = \Omega^{(ji)}$. In the case $r = 0$ the right hand side of (1.28) is equal to

$$\frac{1}{k+2} \sum_{1 \le i < j \le n} \Omega^{(ij)}.$$

Combining with (1.27), we obtain the desired equality. Finally, in the case $r = 1$, applying (1.27) to the right hand side of (1.28) we obtain the statement. $\qquad \square$

By integrating the above infinitesimal result, we obtain the following conformal invariance.

PROPOSITION 1.28. *Let Ψ be a horizontal section of the conformal block bundle $\mathcal{E}_{\lambda_1 \cdots \lambda_n}$. Under a Möbius transformation*

$$w_j = \frac{a z_j + b}{c z_j + d}, \quad 1 \le j \le n,$$
$$a, b, c, d \in \mathbf{C}, \quad ad - bc = 1,$$

Ψ_0 behaves as

$$\Psi_0(z_1, \cdots, z_n) = \prod_{j=1}^{n} (c z_j + d)^{-2 \Delta_{\lambda_j}} \Psi_0(w_1, \cdots, w_n).$$

PROOF. The case $r = -1$ of Proposition 1.27 shows that Ψ_0 is invariant under parallel translations

$$w_j = z_j + c, \quad c \in \mathbf{C}, \quad 1 \le j \le n.$$

In the case $r = 0$, the differential operator

$$\sum_{j=1}^{n} z_j \frac{\partial}{\partial z_j}$$

is the so-called Euler operator and it follows from Proposition 1.27 that with respect to the dilatations

$$w_j = \alpha z_j, \quad \alpha \in \mathbf{C}, \quad 1 \leq j \leq n,$$

Ψ_0 behaves as

$$\Psi_0(w_1, \cdots, w_n) = \alpha^{-\Delta_{\lambda_1} - \cdots - \Delta_{\lambda_n}} \Psi_0(z_1, \cdots, z_n).$$

The Möbius transformations of type

$$f_\epsilon(z) = \frac{z}{-\epsilon z + 1}$$

are called the special conformal transformations. Since for these transformations we have

$$\frac{d}{d\epsilon}\bigg|_{\epsilon=0} \Psi_0\left(f_\epsilon(z_1), \cdots, f_\epsilon(z_n)\right) = \sum_{j=1}^{m} z_j^2 \frac{d}{dz_j} \Psi_0\left(z_1, \cdots, z_n\right),$$

the integral form of the case $r = 1$ of Proposition 1.27 is written as

$$\Psi_0\left(f_\epsilon(z_1), \cdots, f_\epsilon(z_n)\right)$$
$$= \prod_{j=1}^{n} (-\epsilon z_j + 1)^{2\Delta_{\lambda_j}} \Psi_0\left(z_1, \cdots, z_n\right).$$

It is known that the group of Möbius transformations is generated by the above three types of transformations: parallel translations, dilatations and special conformal transformations. Thus we obtain the statement of our proposition.   □

By means of the above conformal invariance we see that the horizontal sections of the conformal block bundle defined over $\mathrm{Conf}_n(\mathbf{C})$ extends to $\mathrm{Conf}_n(\mathbf{C}P^1)$. To summarize we have shown the following important properties of the horizontal sections of the conformal block bundles.

1. Algebraic constraints (Proposition 1.18).
2. KZ equation (Theorem 1.25).
3. Conformal invariance (Proposition 1.28).

1.6. Vertex operators and OPE

The purpose of this section is to reformulate the conformal field theory from the point of view of field operators and the operator product expansions. We fix a global coordinate function z for $\mathbf{C}P^1 = \mathbf{C} \cup \{\infty\}$ as in Section 1.4 . We take $p \in \mathbf{C}P^1$ such that $z(p) \neq 0, \infty$ and set $z = z(p)$. Let us consider the space of conformal blocks for three points: the origin $O \in \mathbf{C}$, p and ∞. With these points we associate level k highest weights λ_0, λ and λ_∞ respectively. We consider integrable highest weight modules of the affine Lie algebra $\widehat{\mathfrak{g}} = \mathfrak{g} \otimes \mathbf{C}((t)) \oplus \mathbf{C}c$. As defined in Section 1.4, the space of conformal blocks $\mathcal{H}(O, p, \infty; \lambda_0, \lambda, \lambda_\infty^*)$ is the space of multilinear maps

$$\Psi : H_{\lambda_0} \times H_\lambda \times H_{\lambda_\infty}^* \to \mathbf{C}$$

invariant under the diagonal action of the meromorphic functions with values in $\mathfrak{g}$ with poles of any order possibly at O, p and ∞.

We consider the conformal block bundle

$$\mathcal{E} = \bigcup_{p \in \mathbf{C} \backslash \{0\}} \mathcal{H}(O, p, \infty; \lambda_0, \lambda, \lambda_\infty^*).$$

Let Ψ be a section of $\mathcal{E}$. We introduce a bilinear map

$$\phi(v, z) : H_{\lambda_0} \otimes H_{\lambda_\infty}^* \to \mathbf{C}$$

determined by

$$\phi(v, z)(u \otimes w) = \Psi(z)(u, v, w)$$

for $u \in H_{\lambda_0}$, $v \in H_\lambda$ and $w \in H_{\lambda_\infty}^*$. We see that $\phi(v, z)$ is linear in $v \in H_\lambda$ and that it depends also on z, the coordinate of the point p. We regard $\phi(v, z)$ as a linear operator from H_{λ_0} to $\overline{H}_{\lambda_\infty} = \prod_{d \geq 0} H_{\lambda_\infty}(d)$. The following proposition describes an important property of $\phi(v, z)$ for $v \in V_\lambda$, where V_λ is identified with $H_{\lambda,0}$ in the direct sum decomposition $H_\lambda = \bigoplus_{d \geq 0} H_\lambda(d)$.

PROPOSITION 1.29. *Let Ψ be a section of the above conformal block bundle $\mathcal{E}$. Then the linear operator $\phi(v, z)$, $v \in V_\lambda$, $z \in \mathbf{C} \backslash \{0\}$, defined by $\phi(v, z)(u \otimes w) = \Psi(z)(u, v, w)$ satisfies the commutation relation*

(1.29) $$[X \otimes t^n, \phi(v, z)] = z^n \phi(Xv, z)$$

for $X \otimes t^n \in \widehat{\mathfrak{g}}$.

PROOF. Consider the meromorphic function $f(z) = X \otimes z^n$, $X \in \mathfrak{g}$, $n \in \mathbf{Z}$. The action of $f(z)$ on H_{λ_0}, V_λ and $H_{\lambda_\infty}^*$ via the Laurent expansions at O, p and ∞ are given respectively by

$$f(z)u = (X \otimes t^n)u, \ u \in H_{\lambda_0},$$
$$f(z)v = z^n \, Xv, \ v \in V_\lambda,$$
$$f(z)w = -w \, (X \otimes t^n), \ w \in H_{\lambda_\infty}^*.$$

Therefore the invariance under the action of $f(z) = X \otimes z^n$ implies

$$\Psi((X \otimes t^n)u, v, w) + z^n \Psi(u, Xv, w) - \Psi(u, v, w(X \otimes t^n)) = 0.$$

The above invariance under the meromorphic function $X \otimes z^n$ is translated into the commutation relation for $X \otimes t^n \in \widehat{\mathfrak{g}}$ and $\phi(v, z)$ expressed as

$$[X \otimes t^n, \phi(v, z)] = z^n \phi(Xv, z).$$

This completes the proof. $\square$

This relation (1.29) is called the *gauge invariance*. As shown in Lemma 1.17, Ψ is uniquely determined by its restriction Ψ_0 on $V_{\lambda_0} \otimes V_\lambda \otimes V_{\lambda_\infty}^*$. Let us now suppose that Ψ is a horizontal section of the vector bundle with respect to the connection ∇. Such an operator

$$\Psi(z) : H_{\lambda_0} \otimes H_\lambda \otimes H_{\lambda_\infty}^* \to \mathbf{C}$$

is called a *chiral vertex operator*. The associated operator $\phi(v, z), v \in V_\lambda$, is called a *primary field*. The operators $\phi(v, z)$, $v \in \bigoplus_{d>0} H_\lambda(d)$, are called *secondary fields* or *descendents*. The following proposition is derived directly from Proposition 1.19.

PROPOSITION 1.30. *A non-trivial chiral vertex operator*

$$\Psi(z) : H_{\lambda_0} \otimes H_\lambda \otimes H_{\lambda_\infty}^* \to \mathbf{C}$$

exists if and only if the highest weights λ_0, λ and λ_∞ satisfy the quantum Clebsch-Gordan condition at level k. In this case the chiral vertex operator is determined up to a constant factor.

We can show by means of the conformal invariance that for the above chiral vertex operator its restriction

$$\Psi_0 : V_{\lambda_0} \otimes V_\lambda \to V_{\lambda_\infty}$$

is written up to a constant factor as

$$z^{\Delta_{\lambda_\infty} - \Delta_{\lambda_0} - \Delta_\lambda} \mathbf{p}$$

where $\mathbf{p}$ is a basis of $\mathrm{Hom}_{\mathfrak{g}}(V_{\lambda_0} \otimes V_\lambda, V_{\lambda_\infty})$. The chiral vertex operator Ψ is uniquely determined by Ψ_0. We decompose $\phi(v, z)$ into homogeneous parts as

$$\phi(v, z) = \sum_{n \in \mathbf{Z}} \phi_n(v, z)$$

where ϕ_n sends $H_\lambda(d)$ to $H_\lambda(d-n)$. By using the above expression for Ψ_0 and the gauge invariance (1.29) one can show that $\phi_0(v, z)$ satisfies the relation

$$[L_0, \phi_0(v, z)] = \left(z\frac{d}{dz} + \Delta_\lambda\right)\phi_0(v, z).$$

It can be shown that the primary field $\phi(v, z), v \in V_\lambda$, is written in the form

$$\phi(v, z) = \sum_{n \in \mathbf{Z}} \phi_n(v) z^{-n-\Delta}$$

where $\Delta = -\Delta_{\lambda_\infty} + \Delta_{\lambda_0} + \Delta_\lambda$ and $\phi_n(v)$ is an operator sending $H_{\lambda_0}(d)$ to $H_{\lambda_\infty}(d-n)$. By some computation using the gauge invariance (1.29) and the definition of the Sugawara operators we obtain the following commutation relation.

PROPOSITION 1.31. *The primary field* $\phi(v, z), v \in V_\lambda$, *satisfies the relation*

$$(1.30) \qquad [L_n, \phi(v, z)] = z^n\left(z\frac{d}{dz} + (n+1)\Delta_\lambda\right)\phi(v, z)$$

for any integer n.

The commutation relation (1.30) can be interpreted as the invariance of

$$\phi(v, z)(dz)^{\Delta_\lambda}$$

under local holomorphic conformal transformations. Let us give an intuitive explanation. The above invariance of $\phi(v, z)(dz)^{\Delta_\lambda}$ under a holomorphic transformation $w = f(z)$ is expressed as

$$\phi(v, f(z)) = \left(\frac{df}{dz}\right)^{-\Delta_\lambda}\phi(v, z).$$

The invariance under a holomorphic transformation $f_\epsilon(z) = z - \epsilon(z)$ implies that the infinitesimal variation $\delta_\epsilon\phi(v, z)$ of $\phi(v, z)$ under the transformation f_ϵ is written as

$$\delta_\epsilon\phi(v, z) = \left(\Delta_\lambda\epsilon'(z) + \epsilon(z)\frac{d}{dz}\right)\phi(v, z).$$

In particular, in the case $\epsilon(z) = \epsilon z^{n+1}$ the above infinitesimal variation coincides with the right hand side of (1.30). On the other hand, since the one-parameter family of transformations $f_\epsilon(z) = z - \epsilon(z)$ generates the vector field $-z^{n+1}\frac{d}{dz}$ the left hand side of (1.30) is the associated infinitesimal transformation.

We define $X(z)$ and $T(z)$ by

$$(1.31) \qquad X(z) = \sum_{n \in \mathbf{Z}} (X \otimes t^n) z^{-n-1},$$

$$(1.32) \qquad T(z) = \sum_{n \in \mathbf{Z}} L_n z^{-n-2}$$

where $X \in \mathfrak{g}$ and both $X(z)$ and $T(z)$ are formal power series in z. For $z \in \mathbf{C} \setminus \{0\}$ we regard $X(z)$ and $T(z)$ as operators from H_λ to its completion $\overline{H}_\lambda = \prod_{d \geq 0} H_\lambda(d)$. Although $X(z)$ is a formal power series with coefficients in $\widehat{\mathfrak{g}}$, for $u \in H_\lambda$ and $\eta \in H_\lambda^*$,

$$\langle \eta, X(z)u \rangle = \sum_{n \in \mathbf{Z}} \langle \eta, (X \otimes t^n) z^{-n-1} u \rangle$$

is expressed as a finite sum. A similar result holds for $T(z)$. The operator $T(z)$ is called the *energy momentum tensor*.

We now introduce the notion of *operator product expansions* (OPE) which describe the behaviour of the composition of operators. We first deal with the composition $X(w)\phi(v,z)$. This composition is formally expressed as

$$\sum_{k \in \mathbf{Z}} w^{-1} z^{-\Delta-k} \sum_{m \in \mathbf{Z}} \left(\frac{z}{w} \right)^m (X \otimes t^m) \phi_{k-m}(v).$$

Let us assume that $|w| > |z| > 0$. By means of the gauge invariance the above expression can be written as

$$\sum_{k \in \mathbf{Z}} w^{-1} z^{-\Delta-k} \sum_{m \geq 0} \left(\frac{z}{w} \right)^m \phi_k(Xv) + R_1(w-z)$$

where $R_1(w-z)$ is regular in the sense that

$$\langle \eta, R_1(w-z)\xi \rangle$$

is holomorphic in $w - z$ for any $\xi \in H_{\lambda_0}$ and $\eta \in H_{\lambda_\infty}^*$.

This shows that in the region $|w| > |z| > 0$ the composition $X(w)\phi(v,z)$ is expressed as

$$(1.33) \qquad X(w)\phi(v,z) = \frac{1}{w-z}\phi(Xu,z) + R_1(w-z)$$

where $R_1(w-z)$ is a regular operator in $w - z$. The composition $X(w)\phi(v,z)$ in the region $|w| > |z| > 0$ is analytically continued to the composition $\phi(v,z)X(w)$ defined in the region $|z| > |w| > 0$. The expression of the composition $X(w)\phi(v,z)$ as in the equation (1.33) is called the operator product expansion of $X(w)$ and $\phi(v,z)$. This describes the singularities of the composition of the operators $X(w)$ and $\phi(v, z)$ along the line $w = z$. As we have seen, the operator product expansion is deduced directly from the commutation relation (1.29). We will explain later that, conversely, the operator product expansion determines the commutation relation for the two operators.

Similarly, for the energy momentum tensor $T(w)$ and the primary field $\phi(v, z)$, we have the operator product expansion

$$(1.34) \quad T(w)\phi(v, z) = \left(\frac{\Delta_\lambda}{(w - z)^2} + \frac{1}{w - z}\frac{\partial}{\partial z} \right) \phi(v, z) + R_2(w - z)$$

in the region $|w| > |z| > 0$, where $R_2(w - z)$ is regular in $w - z$. The composition $T(w)\phi(v, z)$ is analytically continued to $\phi(v, z)T(w)$ defined in the region $|z| > |w| > 0$.

Let us describe the operator product expansion for the composition of the energy momentum tensors. In the region $|w| > |z| > 0$ we have the operator product expansion

$$(1.35)$$
$$T(w)T(z) = \frac{c/2}{(w - z)^4} + \frac{2T(z)}{(w - z)^2} + \frac{1}{w - z}\frac{\partial}{\partial z}T(z) + R_3(w - z)$$

where $R_3(w - z)$ is regular in $w - z$. This is analytically continued to $T(z)T(w)$ defined in the region $|z| > |w| > 0$.

Our next object is to describe how the commutation relations for operators are recovered from the operator product expansions.

LEMMA 1.32. *We have the following expressions for the commutators.*

$$(1.36) \qquad [X \otimes t^n, \phi(v, z)] = \frac{1}{2\pi\sqrt{-1}} \int_C w^n X(w)\phi(v, z)dw,$$

$$(1.37) \qquad [L_n, \phi(v, z)] = \frac{1}{2\pi\sqrt{-1}} \int_C w^{n+1}T(w)\phi(v, z)dw$$

where C is an oriented small circle in the w-plane turning around z counterclockwise.

PROOF. Both equalities can be shown in a similar way by means of the operator product expansions (1.33) and (1.34). We will show the second one. First we fix a point in the w-plane with coordinate z. Let C_1 be an oriented circle in the w-plane with parameter $w = r_1 e^{2\pi\sqrt{-1}\theta}, 0 \leq \theta \leq 1$, where $r_1 > |z|$. It follows from the expression

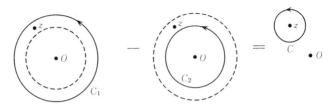

FIGURE 1.3. An interpretation of the commutation relation by contour integrals

$T(w) = \sum_{n \in \mathbf{Z}} L_n w^{-n-2}$ that we have

$$L_n \phi(v, z) = \frac{1}{2\pi\sqrt{-1}} \int_{C_1} w^{n+1} T(w)\phi(v, z) dw.$$

Similarly, we take a circle C_2 with parameter $w = r_2 e^{2\pi\sqrt{-1}\theta}, 0 \le \theta \le 1$, where $r_2 < |z|$. We obtain

$$\phi(v, z) L_n = \frac{1}{2\pi\sqrt{-1}} \int_{C_2} w^{n+1} \phi(v, z) T(w) dw.$$

As we have seen, the operator $T(w)\phi(v, z), |w| > |z| > 0$, is analytically continued to $\phi(v, z)T(w), |z| > |w| > 0$. Therefore, we have

$$\int_{C_1} w^{n+1} T(w)\phi(v, z) dw - \int_{C_2} w^{n+1} \phi(v, z) T(w) dw$$
$$= \int_{C} w^{n+1} T(w)\phi(v, z) dw$$

as shown in Figure 1.3. This completes the proof. $\qquad \Box$

Combining the above lemma and the operator product expansion (1.34), we obtain the commutation relation (1.30). We have thus shown that the commutation relations are dominated by the operator product expansions. The conformal invariance of the primary field $\phi(v, z)$ may be considered to be a consequence of the operator product expansion (1.34). A similar argument can be applied for $X \otimes t^n$ and we see that the gauge invariance (1.29) is deduced from the operator product expansion (1.33).

We now explain that the relations for the Virasoro Lie algebra satisfied by the Sugawara operators can be derived from the operator product expansion (1.35). Let γ be a circle in the z-plane with parameter $z = r e^{2\pi\sqrt{-1}\theta}, 0 \le \theta \le 1$. We can express the Virasoro operator L_n as

$$L_n = \frac{1}{2\pi\sqrt{-1}} \int_\gamma z^{n+1} T(z) dz.$$

We take circles C_1 and C_2 in the w-plane as in the proof of Lemma 1.32 and suppose $r_2 < r < r_1$. We have

$$L_m L_n = \left(\frac{1}{2\pi\sqrt{-1}}\right)^2 \int_\gamma \int_{C_1} w^{m+1} z^{n+1} T(w) T(z) dw dz,$$

$$L_n L_m = \left(\frac{1}{2\pi\sqrt{-1}}\right)^2 \int_\gamma \int_{C_2} w^{m+1} z^{n+1} T(z) T(w) dw dz.$$

Taking a circle C in the w-plane as in Figure 1.3 we have

$$[L_m, L_n] = \left(\frac{1}{2\pi\sqrt{-1}}\right)^2 \int_\gamma \left(\int_C w^{m+1} z^{n+1} T(w) T(z) dw\right) dz.$$

Combining with the operator product expansion (1.35), we derive the commutation relations of the Virasoro Lie algebra in Proposition 1.7.

In order to describe the behaviour of the energy momentum tensor under conformal transformations we introduce the *Schwarzian derivative*

$$(1.38) \qquad S(f, z) = \frac{f'''(z)}{f'(z)} - \frac{3}{2}\left(\frac{f''(z)}{f'(z)}\right)^2$$

for a transformation f of the complex plane. Let us recall the following property of the Schwarzian derivative.

LEMMA 1.33. *For a Möbius transformation*

$$f(z) = \frac{az + b}{cz + d}, \quad a, b, c, d \in \mathbf{C}, ad - bc = 1,$$

the equality $S(f, z) = 0$ holds for any $z \in \mathbf{C}$. Conversely, if $S(f, z) = 0$ for any $z \in \mathbf{C}$, then f is a Möbius transformation.

As in the case of the primary field, it follows from the operator product expansion (1.35) that the infinitesimal variation $\delta_\epsilon T(z)$ with respect to a holomorphic conformal transformation $f_\epsilon(z) = z - \epsilon(z)$ is expressed as

$$(1.39) \qquad \delta_\epsilon T(z) = \epsilon(z)\frac{\partial}{\partial z} T(z) + 2\epsilon'(z) T(z) + \frac{c}{12}\epsilon'''(z).$$

The existence of the last term $\frac{c}{12}\epsilon'''(z)$ shows that $T(z)$ is not a covariant tensor of order 2 with respect to a holomorphic conformal transformation. In the case $\epsilon(z) = z^{n+1}, n = -1, 0, 1, f_\epsilon$ generates a

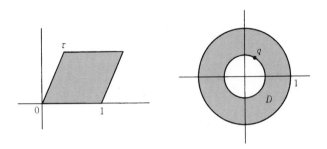

FIGURE 1.4. A fundamental domain of a torus and
its image under the transformation $w = e^{2\pi\sqrt{-1}z}$

global Möbius transformation and the term $\frac{c}{12}\epsilon'''(z)$ vanishes. As an integral form for the equation (1.39) we obtain the following proposition.

PROPOSITION 1.34. *The energy momentum tensor $T(z)$ behaves under a holomorphic conformal transformation $w = f(z)$ as*

$$T(z) = \left(\frac{\partial w}{\partial z}\right)^2 T(w) + \frac{c}{12}S(f, z).$$

Based on the above investigation we quickly review the space of conformal blocks for a torus. Let τ be an element of the upper half plane

$$\mathbf{H} = \{z \in \mathbf{C} \mid \mathrm{Im}\ z > 0\}.$$

The lattice $\mathbf{Z} \oplus \mathbf{Z}\tau$ generated by 1 and τ is denoted by Γ. We shall formulate the space of conformal blocks for $E = \mathbf{C}/\Gamma$. Up to a complex analytic isomorphism we may suppose $0 \leq \mathrm{Re}\ \tau < 1$ and set $q = e^{2\pi\sqrt{-1}\tau}$. We denote by G the transformation group of $\mathbf{C}^* = \mathbf{C} \setminus \{0\}$ generated by the dilatation $f(w) = qw, w \in \mathbf{C}^*$. The orbit space $\mathbf{C}^*/G$ is isomorphic to the torus E under the correspondence $w = e^{2\pi\sqrt{-1}z}$. As depicted in Figure 1.4 the image of the parallelogram spanned by 1 and τ in the z-plane is the annulus D defined by

$$|q| \leq |w| \leq 1$$

in the w-plane. The torus E is obtained from D by identifying the circle $|w| = 1$ and $|w| = |q|$ under the dilatation $f(w) = qw$.

We take a coordinate function $\zeta = 2\pi\sqrt{-1}z$ for the torus $\mathbf{C}^*/G$. Since $w = e^\zeta$ the Schwarzian derivative $S(w, z)$ is equal to $-\frac{1}{2}$. Therefore, by Proposition 1.34 the energy momentum tensor is expressed in terms of the coordinate function as

$$T(\zeta) = \sum_{n \in \mathbf{Z}} \left(L_n - \frac{c}{24}\delta_{n,0} \right) e^{-n\zeta}.$$

It follows that the Virasoro operator L_0' on the torus $\mathbf{C}^*/G$ is expressed as

$$L_0' = L_0 - \frac{1}{24}c.$$

The action of the dilatation $f(w) = qw$ on the integrable highest weight module H_λ is given by

$$(1.40) \qquad q^{L_0 - \frac{1}{24}c}\xi = \exp 2\pi\sqrt{-1}\tau \left(\Delta_\lambda + d - \frac{c}{24} \right) \xi, \quad \xi \in H_\lambda(d),$$

where $H_\lambda = \bigoplus_{d \geq 0} H_\lambda(d)$ is the spectral decomposition in (1.11).

We take distinct n points $p_1, \cdots, p_n$ on the torus $\mathbf{C}^*/G$ and we represent them as points in D. We associate level k highest weights $\mu_1, \cdots, \mu_n$ to $p_1, \cdots, p_n$. In the w-plane we associate H_λ to the origin and its dual module H_λ^* to the infinity. Let Ψ be an element of the space of conformal blocks

$$\mathcal{H}(O, p_1, \cdots, p_n, \infty; \lambda, \mu_1, \cdots, \mu_n, \lambda^*),$$

which is a linear operator

$$\Psi : H_\lambda \otimes H_{\mu_1} \otimes \cdots \otimes H_{\mu_n} \to \overline{H}_\lambda.$$

We formally consider

$$\mathrm{Tr}_{H_\lambda} q^{L_0 - \frac{c}{24}} : H_{\mu_1} \otimes \cdots \otimes H_{\mu_n} \to \mathbf{C}$$

and denote by

$$\mathcal{H}_\lambda(D; p_1, \cdots, p_n; \mu_1, \cdots, \mu_n)$$

the vector space generated by such linear operators $\mathrm{Tr}_{H_\lambda} q^{L_0 - \frac{c}{24}}$ for any

$$\Psi \in \mathcal{H}(O, p_1, \cdots, p_n, \infty; \lambda, \mu_1, \cdots, \mu_n, \lambda^*).$$

We define the space of conformal blocks for the torus E with points $p_1, \cdots, p_n$ and highest weights $\mu_1, \cdots, \mu_n$ by

$$\mathcal{H}(E; p_1, \cdots, p_n; \mu_1, \cdots, \mu_n) = \bigoplus_{0 \leq \lambda \leq k} \mathcal{H}_\lambda(D; p_1, \cdots, p_n; \mu_1, \cdots, \mu_n)$$

where the sum is for any level k highest weight λ.

In particular, in the case $n = 0$, a basis of the space of conformal blocks $\mathcal{H}(E)$ is given by

$$(1.41) \qquad \chi_\lambda(\tau) = \mathrm{Tr}_{H_\lambda} q^{L_0 - \frac{c}{24}}, \quad \lambda = 0, 1, \cdots, k.$$

These are characters of the affine Lie algebra of type $A_1^{(1)}$. The above basis of the space of conformal blocks for the torus is called the *Verlinde basis*. As is shown in [**29**] the characters $\chi_\lambda(\tau), 0 \le \lambda \le k$, are expressed by theta functions and behave under the action of $SL_2(\mathbf{Z})$ as

$$(1.42) \qquad \chi_\lambda\left(-\frac{1}{\tau}\right) = \sum_\mu S_{\lambda\mu}\chi_\mu(\tau),$$

$$(1.43) \qquad \chi_\lambda(\tau + 1) = \exp 2\pi\sqrt{-1}\left(\Delta_\lambda - \frac{c}{24}\right)\chi_\lambda(\tau)$$

where $S_{\lambda\mu}$ and Δ_λ are given by

$$(1.44) \qquad S_{\lambda\mu} = \sqrt{\frac{2}{k+2}}\sin\frac{(\lambda+1)(\mu+1)}{k+2},$$

$$(1.45) \qquad \Delta_\lambda = \frac{\lambda(\lambda+2)}{4(k+2)}.$$

Let us recall that the conformal weight Δ_λ is the eigenvalue of the Sugawara operator L_0 on V_λ. We put $S = (S_{\lambda\mu})$ and define T to be the diagonal matrix with entries

$$\exp 2\pi\sqrt{-1}\left(\Delta_\lambda - \frac{c}{24}\right), \quad 0 \le \lambda \le k.$$

We see that both S and T are symmetric unitary matrices and satisfy the relation

$$S^2 = (ST)^3 = I.$$

Let R_k be a complex vector space with basis $v_\lambda, 0 \le \lambda \le k$. We introduce a product structure by defining

$$v_\lambda \cdot v_\mu = \sum_\nu N_{\lambda\mu}^\nu v_\nu$$

and extending it linearly on R_k. Here $N_{\lambda\mu}^\nu$ is the dimension of the space of conformal blocks for the Riemann sphere with three points

$$\mathcal{H}(p_1, p_2, p_3; \lambda, \mu, \nu^*).$$

Let us recall that $N_{\lambda\mu}^\nu = N_{\lambda\mu\nu}$ holds since we are dealing with $\mathfrak{g} = sl_2(\mathbf{C})$ and that the structure constant $N_{\lambda\mu}^\nu$ is either 0 or 1 as we described in Proposition 1.19.

PROPOSITION 1.35. *The algebra R_k is commutative and associative.*

PROOF. The commutativitiy follows from the symmetry $N_{\lambda\mu}^\nu = N_{\mu\lambda}^\nu$. We shall show the associativity. We have

$$(v_{\lambda_1} \cdot v_{\lambda_2}) \cdot v_{\lambda_3} = \sum_{\lambda,\lambda_4} N_{\lambda_1\lambda_2}^\lambda N_{\lambda\lambda_3}^{\lambda_4} v_{\lambda_4},$$

$$v_{\lambda_1} \cdot (v_{\lambda_2} \cdot v_{\lambda_3}) = \sum_{\mu,\lambda_4} N_{\lambda_1\mu}^{\lambda_4} N_{\lambda_2\lambda_3}^\mu v_{\lambda_4}.$$

By computing the dimension of the space of conformal blocks of the Riemann sphere with four points and highest weights $\lambda_j, 1 \leq j \leq 4$, using the basis as shown in Figure 1.2, we have

$$\sum_\lambda N_{\lambda_1\lambda_2}^\lambda N_{\lambda\lambda_3}^{\lambda_4} = \sum_\mu N_{\lambda_1\mu}^{\lambda_4} N_{\lambda_2\lambda_3}^\mu,$$

which shows the associativity of R_k.   □

The algebra R_k is called the *Verlinde algebra* or the *fusion algebra* for the $SU(2)$ Wess-Zumino-Witten model at level k. The element v_0 is the unit in the algebra R_k. It can be shown that we have an isomorphism

$$\phi : \mathbf{C}[X]/(X^{k+1}) \to R_k$$

between the truncated polynomial ring and the fusion algebra defined by $\phi(X) = v_1$.

PROPOSITION 1.36 (Verlinde formula). *The dimension of the space of conformal blocks*

$$N_{\lambda\mu\nu} = \dim \mathcal{H}(p_1, p_2, p_3; \lambda, \mu, \nu)$$

is given by the formula

$$N_{\lambda\mu\nu} = \sum_\alpha \frac{S_{\lambda\alpha} S_{\mu\alpha} S_{\nu\alpha}}{S_{0\alpha}}$$

using the modular transformation S matrix $S = (S_{\lambda\mu})$.

PROOF. We denote by N_λ the square matrix of degree $k+1$ with $\mu\nu$ entries $N_{\lambda\mu\nu}$, $0 \leq \mu,\nu \leq k$. In particular, in the case $\lambda = 1$, $N_{1\mu\nu} = 1$ if $|\mu - \mu| = 1$ and $N_{1\mu\nu} = 0$ otherwise. The matrix N_1 is diagonalized by the matrix S in (1.44) with eigenvalues $S_{\lambda1}/S_{00}$, $\lambda = 0, 1, \cdots, k$. Since $v_\lambda, \lambda \geq 1$, is expressed as a polynomial in v_1, the matrix $N_\lambda, \lambda \geq 1$, is written as a polynomial in N_1 as well.

It follows that N_λ is diagonalized by S. It can be shown that the eigenvalues of N_λ are $S_{\lambda\mu}/S_{0\mu}$, $\mu = 0, 1, \cdots, k$. Thus we obtain the Verlinde formula. $\qquad\qquad\qquad\qquad\qquad\qquad\qquad\qquad\qquad\qquad\quad$ □

The above proposition may be reformulated in the following way. We take a basis $\{w_\mu\}$ of R_k given by

$$w_\mu = S_{0\mu} \sum_\lambda S_{\lambda\mu} v_\lambda.$$

Then, it follows that

$$w_\lambda \cdot w_\mu = \delta_{\lambda\mu} w_\lambda.$$

In particular, each w_λ, $0 \le \lambda \le k$, is an idempotent. The original assertion due to Verlinde is that the fusion algebra of a rational conformal field theory is diagonalized by the modular transformation S matrix. Proposition 1.36 is a special case.

Let us now describe a basis for the space of conformal blocks of the Riemann sphere with n points $p_1, \cdots, p_n$ and highest weights $\lambda_1, \cdots, \lambda_n$:

$$\mathcal{H}(p_1, \cdots, p_n; \lambda_1, \cdots, \lambda_n).$$

We take level k highest weights μ_j, $0 \le j \le n$, with $\mu_0 = \mu_n = 0$ in such a way that the triplet $(\mu_{j-1}, \lambda_j, \mu_j)$ satisfies the quantum Clebsch-Gordan condition at level k for each j, $1 \le j \le n$. We consider chiral vertex operators

$$\Psi_j(z) : H_{\mu_{j-1}} \otimes H_{\lambda_j} \to \overline{H}_{\mu_j}, \ 1 \le j \le n,$$

and the associated operators

$$\phi_j(z, \xi_j) : H_{\mu_{j-1}} \to \overline{H}_{\mu_j}, \ \xi_j \in H_{\lambda_j}, \ 1 \le j \le n.$$

We denote by $v_0 \in H_0$ a highest weight vector and by v_0^* its dual vector in H_0^*. In the region $0 < |z_1| < |z_2| < \cdots < |z_n|$ the composition

$$\phi_n(z_n, \xi_n) \cdots \phi_1(z_1, \xi_1) : H_0 \to \overline{H}_0$$

is defined and the correspondence

$$\xi_1 \otimes \cdots \otimes \xi_n \mapsto \langle v_0^*, \phi_n(z_n, \xi_n) \cdots \phi_1(z_1, \xi_1) v_0 \rangle$$

determines a multilinear map

$$\Psi_{\mu_0 \mu_1 \cdots \mu_n}(z_1, \cdots, z_n) : H_{\lambda_1} \otimes \cdots \otimes H_{\lambda_n} \to \mathbf{C}.$$

It can be shown that its restriction on $V_{\lambda_1} \otimes \cdots \otimes V_{\lambda_n}$ satisfies the KZ equation and the algebraic equations in Section 1.5. The operators $\Psi_{\mu_0 \mu_1 \cdots \mu_n}(z_1, \cdots, z_n)$ for any $(\mu_0, \mu_1, \cdots, \mu_n)$ satisfying the above quantum Clebsch-Gordan condition at level k form a basis of our space of conformal blocks. We refer the reader to Tsuchiya and Kanie [51] for details. We have the following dimension formula for the space of conformal blocks.

PROPOSITION 1.37. *The dimension of the space of conformal blocks for the Riemann sphere with n points is given by*

$$\dim \mathcal{H}(p_1, \cdots, p_n; \lambda_1, \cdots, \lambda_n) = \sum_{0 \leq \lambda \leq k} \frac{S_{\lambda_1 \lambda} \cdots S_{\lambda_n \lambda}}{(S_{0\lambda})^{n-2}}.$$

PROOF. Let us suppose $n \geq 3$. By the above description of a basis of the space of conformal blocks

$$\dim \mathcal{H}(p_1, \cdots, p_n; \lambda_1, \cdots, \lambda_n) = \sum_{\mu_1, \cdots, \mu_{n-1}} \prod_{j=1}^{n} N_{\mu_{j-1} \lambda_j \mu_j}.$$

Applying Proposition 1.36 and the fact that S is a symmetric and unitary matrix, we obtain the desired formula. The case $n = 1, 2$ can be verified directly since we have $\dim \mathcal{H}(p; \lambda) = \delta_{\lambda 0}$ and

$$\dim \mathcal{H}(p_1, p_2; \lambda_1, \lambda_2) = \delta_{\lambda_1 \lambda_2}$$

as shown in Section 1.4.   □

CHAPTER 2

Jones-Witten Theory

In Section 2.1, we will study the structure of the holonomy of the KZ connection on the conformal block bundle. As the holonomy we obtain linear representations of the braid groups. The Jones polynomial was first defined in the framework of the theory of operator algebras. Later, a combinatorial approach for the Jones polynomial was developed, especially using the theory of quantum groups. In this book we will construct the Jones polynomial directly from the holonomy of the conformal block bundle. This geometric approach to the Jones polynomial will be developed in Section 2.2. In Section 2.3, we will deal with Witten's invariants for 3-manifolds based on conformal field theory. Let us recall that Witten's invariants were originally formulated as the partition function of the Chern-Simons functional on the space of connections over a 3-manifold. In Section 2.4, we will study the action of the mapping class groups on the space of conformal blocks and show that the conformal field theory provides an example of topological quantum field theory. Finally, in Section 2.5, we will describe a relation between the Chern-Simons theory and complex line bundles on the space of connections over a Riemann surface. We will sketch briefly how the Chern-Simons gauge theory is related to the conformal field theory.

2.1. KZ equation and representations of braid groups

We denote by H_{ij} the hyperplane in the complex vector space $\mathbf{C}^n$ defined by $z_i = z_j$, where $z_1, \cdots, z_n$ are coordinate functions for $\mathbf{C}^n$. Let us recall that the configuration space of n ordered distinct points in the complex plane $\mathbf{C}$ is by definition

$$\mathrm{Conf}_n(\mathbf{C}) = \mathbf{C}^n \setminus \bigcup_{i<j} H_{ij}.$$

The fundamental group of the configuration space $\mathrm{Conf}_n(\mathbf{C})$ is called the *pure braid group* on n strands. The symmetric group $\mathcal{S}_n$ acts on

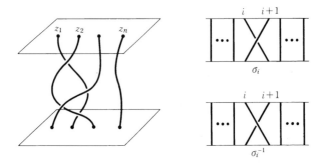

FIGURE 2.1. The braid group and its generators

$\mathrm{Conf}_n(\mathbf{C})$ by permutations of coordinates. The fundamental group of the quotient space $Y_n = \mathrm{Conf}_n(\mathbf{C})/\mathcal{S}_n$ by the above action of $\mathcal{S}_n$ is called the *braid group* on n strands and is denoted by B_n. We have a natural projection map $p : \mathrm{Conf}_n(\mathbf{C}) \to Y_n$, which is a Galois covering map. Let $x_0 = (a_1, \cdots, a_n)$ be a base point in $\mathrm{Conf}_n(\mathbf{C})$. We choose x_0 in such a way that $a_1, \cdots, a_n \in \mathbf{R}$ and $a_1 < \cdots < a_n$. The map p induces an injective homomorphism of fundamental groups

$$p_* : \pi_1(\mathrm{Conf}_n(\mathbf{C}), x_0) \to \pi_1(Y_n, p(x_0))$$

whose cokernel is the covering transformation group $\mathcal{S}_n$. We have the exact sequence

$$1 \to P_n \to B_n \to \mathcal{S}_n \to 1.$$

Each element of B_n is represented by an isotopy class of disjoint n strands in $\mathbf{C} \times [0, 1]$ connecting $\{a_1, \cdots, a_n\} \times \{0\}$ and $\{a_1, \cdots, a_n\} \times \{1\}$ as depicted in Figure 2.1. It is known by E. Artin [4] that B_n is generated by braids σ_i, $1 \le i \le n-1$, as shown in Figure 2.1 with relations

(2.1) $\qquad \sigma_i \sigma_{i+1} \sigma_i = \sigma_{i+1} \sigma_i \sigma_{i+1}, \quad i = 1, \cdots, n-2,$

(2.2) $\qquad \sigma_i \sigma_j = \sigma_j \sigma_i, \quad |i - j| > 1.$

See [14] for details.

We consider logarithmic differential 1-forms

$$\omega_{ij} = d\log(z_i - z_j)$$
$$= \frac{dz_i - dz_j}{z_i - z_j}, \quad i \ne j,$$

defined on $\mathrm{Conf}_n(\mathbf{C})$. These differential forms satisfy the quadratic relations

$$(2.3) \quad \omega_{ij} \wedge \omega_{jk} + \omega_{jk} \wedge \omega_{ik} + \omega_{ik} \wedge \omega_{ij} = 0, \quad i < j < k,$$

called the *Arnold relations*. These relations were used by Arnold [2] to give a presentation of the cohomology ring $H^*(\mathrm{Conf}_n(\mathbf{C}), \mathbf{Z})$. Let us recall it briefly. The cohomology ring $H^*(\mathrm{Conf}_n(\mathbf{C}), \mathbf{Z})$ is isomorphic to the exterior algebra generated by $e_{ij}, 1 \leq i < j \leq n$, with degree 1 modulo the ideal generated by

$$e_{ij} \wedge e_{jk} + e_{jk} \wedge e_{ik} + e_{ik} \wedge e_{ij}, \quad i < j < k.$$

Here e_{ij} is represented by the de Rham cohomology class of

$$\frac{1}{2\pi\sqrt{-1}} \, \omega_{ij}.$$

In Section 1.5, we introduced the KZ connection as a natural flat connection on the conformal block bundle. Let us again formulate the KZ equation, in general, for a finite dimensional complex semisimple Lie algebra $\mathfrak{g}$. We fix a finite dimensional complex semisimple Lie algebra $\mathfrak{g}$ and its representations $\rho_j : \mathfrak{g} \to \mathrm{End}(V_j), 1 \leq j \leq n$. We denote by $\{I_\mu\}$ an orthonormal basis of $\mathfrak{g}$ with respect to the Cartan-Killing form and set $\Omega = \sum_\mu I_\mu \otimes I_\mu$. For example, in the case $\mathfrak{g} = sl_2(\mathbf{C})$, Ω may be written as

$$\Omega = \frac{1}{2} H \otimes H + E \otimes F + F \otimes E.$$

The element $C = \sum_\mu I_\mu I_\mu$ in the universal enveloping algebra $U(\mathfrak{g})$ is called the *Casimir element*. The comultiplication

$$\Delta : U(\mathfrak{g}) \to U(\mathfrak{g}) \otimes U(\mathfrak{g})$$

is the homomorphism of algebras characterized by $\Delta(X) = X \otimes 1 + 1 \otimes X$ for any $X \in \mathfrak{g}$. We have

$$(2.4) \qquad \Omega = \frac{1}{2}(\Delta C - C \otimes 1 - 1 \otimes C).$$

Let $\phi : V_1 \otimes V_2 \otimes \cdots \otimes V_n \to \mathbf{C}$ be a multilinear form. We denote by $\Omega^{(ij)}\phi$ the multilinear form defined by

$$(2.5) \quad (\Omega^{(ij)}\phi)(v_1 \otimes \cdots \otimes v_n)$$
$$= \sum_\mu \phi(v_1 \otimes \cdots \otimes \rho_i(I_\mu)v_i \otimes \cdots \otimes \rho_j(I_\mu)v_j \otimes \cdots \otimes v_n)$$

for $v_1 \otimes \cdots \otimes v_n \in V_1 \otimes V_2 \otimes \cdots \otimes V_n$. The KZ equation is by definition a system of linear partial differential equations for a function $\Phi(z_1, \cdots, z_n)$ defined over $\mathrm{Conf}_n(\mathbf{C})$ with values in

$$\mathrm{Hom}_{\mathbf{C}}(V_1 \otimes V_2 \otimes \cdots \otimes V_n, \mathbf{C})$$

given as

$$(2.6) \qquad \qquad \frac{\partial \Phi}{\partial z_i} = \frac{1}{\kappa} \sum_{j, j \neq i} \frac{\Omega^{(ij)} \Phi}{z_i - z_j}$$

where κ is a non-zero complex parameter. By putting

$$(2.7) \qquad \qquad \omega = \frac{1}{\kappa} \sum_{1 \leq i < j \leq n} \Omega^{(ij)} \omega_{ij}$$

the KZ equation is also written in the form of a total differential equation

$$(2.8) \qquad \qquad d\Phi = \omega \Phi.$$

LEMMA 2.1. *The above* $\Omega^{(ij)}, 1 \leq i \neq j \leq n$, *satisfy the following relations:*

1. $\Omega^{(ij)} = \Omega^{(ji)}$,
2. $[\Omega^{(ij)} + \Omega^{(jk)}, \Omega^{(ik)}] = 0$, $\quad i, j, k \quad distinct$,
3. $[\Omega^{(ij)}, \Omega^{(kl)}] = 0$, $\quad i, j, k, l \quad distinct$.

PROOF. The relations 1 and 3 are clear. We shall show the relation 2. It is enough to consider the case $n = 3$. Let us recall that the Casimir element C lies in the center of the universal enveloping algebra $U(\mathfrak{g})$. Thus,

$$[\Delta(C), \Delta(X)] = 0$$

in $U(\mathfrak{g}) \otimes U(\mathfrak{g})$ for any $X \in U(\mathfrak{g})$. Thus, we obtain

$$[\Delta(C) \otimes 1, \sum_\mu \Delta(I_\mu) \otimes I_\mu] = 0.$$

Together with the equation (2.4), we have

$$[\Omega^{(12)}, \Omega^{(13)} + \Omega^{(23)}] = 0$$

since $C \otimes 1 \otimes 1$ and $1 \otimes C \otimes 1$ lie in the center of $U(\mathfrak{g}) \otimes U(\mathfrak{g}) \otimes U(\mathfrak{g})$. The other relations are obtained in a similar way. This completes the proof. $\qquad \square$

Combining the above lemma with the Arnold relation (2.3), we have the following relation for ω defined in (2.7).

LEMMA 2.2. *We have*

$$\omega \wedge \omega = 0.$$

PROOF. The wedge product $\omega \wedge \omega$ is computed as

$$\omega \wedge \omega = \frac{1}{\kappa^2} \sum_{i<j,k<l} [\Omega^{(ij)}, \Omega^{(kl)}] \omega_{ij} \wedge \omega_{kl}.$$

By means of the Arnold relation we obtain

$$\sum_{i<j,k<l} [\Omega^{(ij)}, \Omega^{(kl)}] \omega_{ij} \wedge \omega_{kl}$$

$$= \sum_{i<j<k} \Big([\Omega^{(ij)} + \Omega^{(jk)}, \Omega^{(ik)}] \omega_{ij} \wedge \omega_{ik}$$

$$+ [\Omega^{(ij)} + \Omega^{(ik)}, \Omega^{(jk)}] \omega_{ij} \wedge \omega_{jk} \Big)$$

$$+ \sum_{\{i,j\} \cap \{k,l\} = \emptyset} [\Omega^{(ij)}, \Omega^{(kl)}] \omega_{ij} \wedge \omega_{kl},$$

which vanishes by Lemma 2.1.   □

Let E denote a trivial vector bundle over the configuration space $\mathrm{Conf}_n(\mathbf{C})$ with fibre

$$(V_1 \otimes V_2 \otimes \cdots \otimes V_n)^* = \mathrm{Hom}_{\mathbf{C}}(V_1 \otimes V_2 \otimes \cdots \otimes V_n, \mathbf{C}).$$

We fix a global trivialization of E and we identify the space of sections of E with the set of functions on $\mathrm{Conf}_n(\mathbf{C})$ with values in $V_1^* \otimes V_2^* \otimes \cdots \otimes V_n^*$. We regard ω in (2.7) as a 1-form on $\mathrm{Conf}_n(\mathbf{C})$ with values in $\mathrm{End}(V_1^* \otimes V_2^* \otimes \cdots \otimes V_n^*)$. We define a connection on E by $\nabla = d - \omega$. The connection ∇ is called the KZ connection. Since $d\omega = 0$ and $\omega \wedge \omega = 0$ by Lemma 2.2, we see that the curvature $d\omega + \omega \wedge \omega$ of the connection ∇ is zero. Therefore, ∇ is a flat connection.

Since a horizontal section of the connection ∇ is a solution of the KZ equation, the holonomy of the connection ∇ can be described in the following way. Let γ be a loop in $\mathrm{Conf}_n(\mathbf{C})$ with a base point x_0. On a small neighbourhood of x_0 we take linearly independent solutions of the KZ equation $(\Phi_1, \cdots, \Phi_m)$, where $m = \dim V_1 \times \cdots \times \dim V_n$. By the analytic continuation along the loop γ, the system of solutions $(\Phi_1, \cdots, \Phi_m)$ is transformed as

$$(\Phi_1, \cdots, \Phi_m) \theta(\gamma)$$

by a matrix $\theta(\gamma)$. The linear transformation $\theta(\gamma)$ depends only on the homotopy class of the loop γ since ∇ is a flat connection. Therefore, we obtain a linear representation of the pure braid group

$$\theta : P_n \to GL(V_1^* \otimes V_2^* \otimes \cdots \otimes V_n^*)$$

with a parameter κ. This is called the *monodromy representation* of the KZ equation. We have $\theta(\sigma\tau) = \theta(\sigma)\theta(\tau)$ for any $\sigma, \tau \in P_n$. Here we adopt a convention that the monodromy representation is a right action on the vector space $V_1^* \otimes V_2^* \otimes \cdots \otimes V_n^*$. To summarize we have the following proposition.

PROPOSITION 2.3. *For any complex semisimple Lie algebra $\mathfrak{g}$ and its representations $\rho_j : \mathfrak{g} \to \mathrm{End}(V_j), 1 \le j \le n$, the holonomy of the KZ connection ∇ gives a linear representation of the pure braid group*

$$\theta : P_n \to GL(V_1^* \otimes V_2^* \otimes \cdots \otimes V_n^*)$$

with a parameter κ.

REMARK 2.4. As we have seen in the proof of Lemma 2.1, we have $[\Delta(C), \Delta(X)] = 0$ for any $X \in U(\mathfrak{g})$. It follows that the KZ connection ∇ is invariant under the diagonal action of the Lie algebra $\mathfrak{g}$. Therefore, the monodromy representation θ leaves invariant the space of coinvariant tensors $\mathrm{Hom}_{\mathfrak{g}}(V_1 \otimes \cdots \otimes V_n, \mathbf{C})$.

In the case $V_1 = \cdots = V_n = V$, the symmetric group $\mathcal{S}_n$ acts diagonally on the total space $\mathrm{Conf}_n(\mathbf{C}) \times (V^{\otimes n})^*$ of the vector bundle E, where the right action of $\mathcal{S}_n$ on $(V^{\otimes n})^* \cong (V^*)^{\otimes n}$ is defined by

$$(\phi \cdot \sigma)(v_1 \otimes \cdots \otimes v_n) = \phi(v_{\sigma(1)} \otimes \cdots \otimes v_{\sigma(n)})$$

for $\phi \in (V^{\otimes n})^*$, $\sigma \in \mathcal{S}_n$ and $v_j \in V_j, 1 \le j \le n$. The quotient space

$$F = \mathrm{Conf}_n(\mathbf{C}) \times_{\mathcal{S}_n} (V^*)^{\otimes n}$$

defines a vector bundle over $Y_n = \mathrm{Conf}_n(\mathbf{C})/\mathcal{S}_n$ with fibre $(V^*)^{\otimes n}$. The KZ connection ∇ is invariant under the above action of $\mathcal{S}_n$. Hence ∇ defines a flat connection on the vector bundle F and the monodromy representation in Proposition 2.3 can be extended to a linear representation of the braid group

$$\theta : B_n \to GL((V^*)^{\otimes n}).$$

Let us deal with the situation of the space of conformal blocks. As in Section 1.4 we are going to investigate the space of conformal blocks of the Riemann sphere for $\mathfrak{g} = sl_2(\mathbf{C})$ at level k, where k is a positive integer. Take four distinct points p_1, p_2, p_3 and p_4 on $\mathbf{CP}^1$ with $p_4 = \infty$. Fix a global coordinate function z as in Section 1.4 and set $z(p_j) = z_j, j = 1, 2, 3$. We associate to the points p_1, p_2, p_3 and p_4

level k highest weights $\lambda_1, \lambda_2, \lambda_3$ and λ_4^* respectively and consider the space of conformal blocks $\mathcal{H}(p_1, p_2, p_3, p_4; \lambda_1, \lambda_2, \lambda_3, \lambda_4^*)$. As we have shown in Lemma 1.17 this space of conformal blocks can be embedded in the space of coinvariant tensors

$$(2.9) \qquad \operatorname{Hom}_{\mathfrak{g}}\left(V_{\lambda_1} \otimes V_{\lambda_2} \otimes V_{\lambda_3} \otimes V_{\lambda_4}^*, \mathbf{C} \right).$$

The associated conformal block bundle $\mathcal{E}$ over $\operatorname{Conf}_3(\mathbf{C})$ admits a flat connection ∇ which is expressed as the KZ connection with $\kappa = k+2$. Since the KZ connection is invariant under parallel translations, we introduce coordinates (ζ_1, ζ_2) given by $\zeta_1 = z_2 - z_1$ and $\zeta_2 = z_3 - z_1$. We perform a coordinate transformation

$$\zeta_1 = u_1 u_2, \quad \zeta_2 = u_2$$

to blow up the triple point defined by $z_1 = z_2 = z_3$. The KZ connection form $\omega = \frac{1}{\kappa} \sum_{1 \le i < j \le 3} \Omega^{(ij)} \omega_{ij}$ is written in the coordinates (u_1, u_2) as

$$(2.10) \qquad \omega = \frac{1}{\kappa} \left(\frac{\Omega^{(12)}}{u_1} du_1 + \frac{\Omega^{(12)} + \Omega^{(13)} + \Omega^{(23)}}{u_2} du_2 + \omega_1 \right)$$

where ω_1 is a holomorphic 1-form around $u_1 = u_2 = 0$. The residues of the connection form ω along $u_1 = 0$ and $u_2 = 0$ are

$$\operatorname{Res}_{u_1=0} \omega = \frac{1}{\kappa} \Omega^{(12)},$$

$$\operatorname{Res}_{u_2=0} \omega = \frac{1}{\kappa} (\Omega^{(12)} + \Omega^{(13)} + \Omega^{(23)}).$$

Let us recall that the space of conformal blocks has a basis which is in one-to-one correspondence with labelled trees as depicted in Figure 1.2 in Section 1.4. We fix a basis $\{\mathbf{p}_\lambda\}$ of the image of the space of conformal blocks in the space of coinvariant tensors (2.9). Here $\mathbf{p}_\lambda$ corresponds to the tree in Figure 1.2 in Section 1.4 with label λ for the internal edge and we suppose that each triple $(\lambda_1, \lambda_2, \lambda)$ and $(\lambda, \lambda_3, \lambda_4)$ satisfies the quantum Clebsch-Gordan condition at level k. The above residue matrices are diagonalized simultaneously for the basis $\{\mathbf{p}_\lambda\}$ with eigenvalues $\Delta_\lambda - \Delta_{\lambda_1} - \Delta_{\lambda_2}$ and $\Delta_{\lambda_4} - \Delta_{\lambda_1} - \Delta_{\lambda_2} - \Delta_{\lambda_3}$ respectively.

PROPOSITION 2.5. *A basis of the space of horizontal sections of the conformal block bundle $\mathcal{E}$ is written around $u_1 = u_2 = 0$ as*

$$u_1^{\Delta_\lambda - \Delta_{\lambda_1} - \Delta_{\lambda_2}} u_2^{\Delta_{\lambda_4} - \Delta_{\lambda_1} - \Delta_{\lambda_2} - \Delta_{\lambda_3}} h_\lambda(u_1, u_2) \mathbf{p}_\lambda$$

for any λ such that each triple $(\lambda_1, \lambda_2, \lambda)$ and $(\lambda, \lambda_3, \lambda_4)$ satisfies the quantum Clebsch-Gordan condition at level k. Here $h_\lambda(u_1, u_2)$ is a single-valued holomorphic function around $u_1 = u_2 = 0$

PROOF. For λ satisfying the above condition, we take chiral vertex operators

$$\Psi^\lambda_{\lambda_1 \lambda_2}(\zeta_1) : H_{\lambda_1} \otimes H_{\lambda_2} \to \overline{H}_\lambda,$$

$$\Psi^{\lambda_4}_{\lambda \lambda_3}(\zeta_2) : H_\lambda \otimes H_{\lambda_3} \to \overline{H}_{\lambda_4}.$$

The composition $\Psi^{\lambda_4}_{\lambda \lambda_3}(\zeta_2)\left(\Psi^\lambda_{\lambda_1 \lambda_2}(\zeta_1) \otimes id_{H_{\lambda_3}}\right)$ defined in the region $|\zeta_2| > |\zeta_1| > 0$ is written as

$$u_1^{\Delta_\lambda - \Delta_{\lambda_1} - \Delta_{\lambda_2}} u_2^{\Delta_{\lambda_4} - \Delta_{\lambda_1} - \Delta_{\lambda_2} - \Delta_{\lambda_3}} h_\lambda(u_1, u_2)\mathbf{p}_\lambda$$

when restricted to the space of coinvariant tensors given in (2.9), where $h_\lambda(u_1, u_2)$ is a power series. Since a horizontal section of $\mathcal{E}$ satisfies the KZ equation it follows that $h_\lambda(u_1, u_2)$ is holomorphic. This completes the proof.   □

It follows from the above proposition that the KZ equation has a system of fundamental solutions around $u_1 = u_2 = 0$ written as

$$(2.11) \qquad \Phi_1(u_1, u_2) = \varphi_1(u_1, u_2) u_1^{\frac{1}{\kappa}\Omega_{12}} u_2^{\frac{1}{\kappa}(\Omega_{12} + \Omega_{13} + \Omega_{23})}$$

with a matrix valued holomorphic function $\varphi_1(u_1, u_2)$. Moreover, $\Phi_1(u_1, u_2)$ is diagonalized with respect to the basis $\{\mathbf{p}_\lambda\}$.

In a similar way, we can construct horizontal sections of $\mathcal{E}$ associated with another tree in Figure 1.2 in Section 1.4 with label μ for the internal edge. We perform a coordinate transformation

$$\zeta_2 - \zeta_1 = v_1 v_2, \quad \zeta_2 = v_2$$

to see that the connection matrix ω is expressed around $v_1 = v_2 = 0$ as

$$(2.12) \qquad \omega = \frac{1}{\kappa}\left(\frac{\Omega^{(23)}}{v_1}dv_1 + \frac{\Omega^{(12)} + \Omega^{(13)} + \Omega^{(23)}}{v_2}dv_2 + \omega_2\right)$$

where ω_2 is a holomorphic 1-form around $v_1 = v_2 = 0$. In this case $\Omega^{(23)}$ and $\Omega^{(12)} + \Omega^{(13)} + \Omega^{(23)}$ are diagonalized simultaneously with respect to the basis associated with the above tree. The KZ equation has a system of solutions around $v_1 = v_2 = 0$ written as

$$(2.13) \qquad \Phi_2(v_1, v_2) = \varphi_2(v_1, v_2) v_1^{\frac{1}{\kappa}\Omega^{(23)}} v_2^{\frac{1}{\kappa}(\Omega^{(12)} + \Omega^{(13)} + \Omega^{(23)})}$$

where $\varphi_2(v_1, v_2)$ is holomorphic around $v_1 = v_2 = 0$.

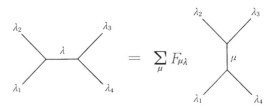

FIGURE 2.2. Graphical expression of the connection matrix

Thus, we have constructed two systems of solutions of the KZ equation, Φ_1 and Φ_2, one normalized around $u_1 = u_2 = 0$ and the other around $v_1 = v_2 = 0$. Suppose p_1, p_2 and p_3 are on the real line and $z_1 < z_2 < z_3$. We consider the asymptotic regions $z_1 < z_2 \ll z_3$ with the solution Φ_1 and $z_1 \ll z_2 < z_3$ with Φ_2. By the analytic continuation along a path from $z_1 < z_2 \ll z_3$ to $z_1 \ll z_2 < z_3$ in the real region, we obtain a constant matrix F such that

$$(2.14) \qquad \Phi_1 = \Phi_2 F.$$

The matrix F is called the *connection matrix*. Let us observe that the points $u_1 = u_2 = 0$ and $v_1 = v_2 = 0$ are both double points of the normal crossing divisors obtained by blowing up the triple point $z_1 = z_2 = z_3$. The connection matrix F relates two solutions of the KZ equation normalized around these double points. The relation between Φ_1 and Φ_2 obtained in this way will be depicted graphically as in Figure 2.2. We represent the two graphs as symbols $(((12)3)4)$ and $((1(23))4)$ respectively.

In terms of the composition of chiral vertex operators the connection matrix is interpreted in the following way.

LEMMA 2.6. *In the region $0 < |\zeta_1| < |\zeta_2|$ we have*

$$\Psi^{\lambda_4}_{\lambda\lambda_3}(\zeta_2) \left(\Psi^{\lambda}_{\lambda_1\lambda_2}(\zeta_1) \otimes id_{H_{\lambda_3}} \right)$$
$$= \sum_\mu F_{\mu\lambda} \Psi^{\lambda_4}_{\lambda_1\mu}(\zeta_2) \left(id_{H_{\lambda_1}} \otimes \Psi^{\mu}_{\lambda_2\lambda_3}(\zeta_2 - \zeta_1) \right).$$

PROOF. The above equality becomes $\Phi_1 = \Phi_2 F$ when we restrict to the space of conformal blocks to the space coinvariant tensors (2.9). Since this restriction map is injective as shown in Lemma 1.17, we obtain the desired equality. $\qquad\square$

By means of the conformal invariance of the solutions of the KZ equation, the above connection matrix F can be described in terms of a Fuchsian differential equation in one variable with regular singularities at 0, 1 and ∞ in the following manner. In the case $n = 3$, the KZ equation has a solution of the form

$$(2.15) \qquad \Phi(z_1, z_2, z_3) = (z_3 - z_1)^{\frac{1}{\kappa}(\Omega^{(12)} + \Omega^{(13)} + \Omega^{(23)})} G\left(\frac{z_2 - z_1}{z_3 - z_1}\right)$$

and G satisfies the differential equation

$$(2.16) \qquad G'(x) = \frac{1}{\kappa}\left(\frac{\Omega^{(12)}}{x} + \frac{\Omega^{(23)}}{x - 1}\right) G(x).$$

We take solutions of the differential equation (2.16) in the open interval $(0, 1)$ with the asymptotic behaviour around $x = 0$ and $x = 1$ described as

$$G_1(x) = H_1(x) x^{\frac{1}{\kappa}\Omega_{12}} \quad (x \to 0),$$
$$G_2(x) = H_2(x)(1 - x)^{\frac{1}{\kappa}\Omega_{23}} \quad (x \to 1)$$

where H_1 and H_2 are holomorphic around $x = 0$ and $x = 1$ respectively. Since the solutions G_1 and G_2 correspond to the solutions Φ_1 and Φ_2 constructed above, we see by the analytic continuation that G_1 and G_2 are related by the connection matrix F introduced in (2.14) as

$$(2.17) \qquad G_1(x) = G_2(x)F.$$

In the case when the parameter κ is generic, the matrix F is expressed as a power series in $\Omega^{(12)}$ and $\Omega^{(23)}$. A universal expression for such power series in $\Omega^{(12)}$ and $\Omega^{(23)}$ is called the Drinfel'd associator. We will discuss it extensively in Section 3.1.

Let Γ_n be a trivalent tree with $n + 1$ external edges. Fix an ordering of the external edges and label them with level k highest weights $\lambda_1, \cdots, \lambda_{n+1}$. Give an orientation to the edges as shown in Figure 2.3. We label each internal edge by a level k highest weight in such a way that the quantum Clebsch-Gordan condition at level k is satisfied for highest weights of edges meeting at any trivalent vertex. Such labelling of internal edges by highest weights is called an admissible labelling. We take $n + 1$ points $p_1, \cdots, p_n, p_{n+1}$ on the Riemann sphere with $p_{n+1} = \infty$ and set $z(p_j) = z_j$. As is explained in Section 1.4 the space of conformal blocks

$$\mathcal{H}(p_1, \cdots, p_n, p_{n+1}; \lambda_1, \cdots, \lambda_n, \lambda_{n+1}^*)$$

has a basis which is in one-to-one correspondence with the set of admissible labelling.

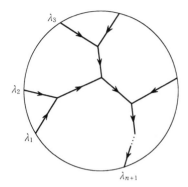

FIGURE 2.3. Tree with $n + 1$ external edges

As in the case $n = 3$ we consider the composition of the chiral vertex operators for each admissible labelling, which gives a basis of the space of conformal blocks. Each trivalent tree represents a system of solutions of the KZ equation. For example, consider a trivalent tree of type $(\cdots (12)3) \cdots n) n + 1)$ having $n + 1$ external edges with highest weights $\lambda_1, \cdots, \lambda_{n+1}$. By performing coordinate transformations $\zeta_k = z_{k+1} - z_1$ and

$$\zeta_k = u_k u_{k+1} \cdots u_{n-1}, \quad k = 1, \cdots, n - 1,$$

we have a system of solutions of the KZ equation around $u_1 = \cdots = u_{n-1} = 0$ of the form

(2.18)
$$\Phi_1 = \varphi_1(u_1, \cdots, u_{n-1}) u_1^{\frac{1}{\kappa}\Omega^{(12)}} u_2^{\frac{1}{\kappa}(\Omega^{(12)}+\Omega^{(13)}+\Omega^{(23)})} \cdots u_{n-1}^{\frac{1}{\kappa}\sum_{1 \le i < j \le n}\Omega^{(ij)}}$$

where $\varphi_1(u_1, \cdots, u_{n-1})$ is a matrix valued holomorphic function. Here Φ_1 is diagonalized with respect to the basis of the space of conformal blocks associated with the admissible labelling of the tree of type $(\cdots (12)3) \cdots n) n + 1)$.

Let us consider the case $n = 4$. There are five types of trees as shown in Figure 2.4 and each tree is associated with a system of solutions of the KZ equation. For the tree of type $(((12)(34))5)$ we perform a coordinate transformation $\zeta_1 = v_1 v_3$, $\zeta_3 - \zeta_2 = v_2 v_3$, $\zeta_3 = v_3$ and associate

(2.19) $$\Phi_2 = \varphi_2(v_1, v_2, v_3) v_1^{\frac{1}{\kappa}\Omega^{(12)}} v_2^{\frac{1}{\kappa}\Omega^{(34)}} v_3^{\frac{1}{\kappa}\sum_{1 \le i < j \le 4}\Omega^{(ij)}}$$

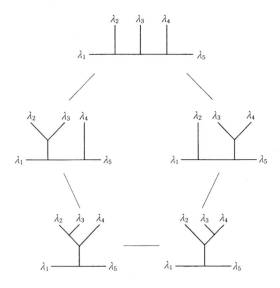

FIGURE 2.4. Pentagon relation

where $\varphi_2(v_1, v_2, v_3)$ is holomorphic around $v_1 = v_2 = v_3 = 0$. It is clear by Lemma 2.6 that the solution Φ_1 in (2.18) in the case $n = 4$ is related to the above Φ_2 by the connection matrix F defined in the case $n = 3$. In general, the connection matrix for each edge path in Figure 2.4 is represented by the composition of connection matrices in the case $n = 3$. A connection matrix in the case $n = 3$ will be called an *elementary connection matrix* and the corresponding linear map is called an *elementary fusing operation*. We have the following *pentagon relation*.

PROPOSITION 2.7. *For the two edge paths connecting two distinct trees in Figure 2.4, the corresponding compositions of the elementary connection matrices coincide.*

PROOF. Let us observe that by blowing up, the region $0 < \zeta_1 < \zeta_2 < \zeta_3 < \zeta_4$ has a compactification as a manifold with corners. It is a contractible 3-dimensional cell complex with five vertices corresponding to the five types of trees in Figure 2.4. Since any path connecting two vertices are homotopic, we obtain the desired result by the flatness of the KZ connection. □

Let us deal with the monodromy representations of the braid groups on the space of conformal blocks. We take n distinct points $p_1, p_2, \cdots, p_n$ with coordinates $z_1, z_2, \cdots, z_n$ satisfying $0 < z_1 < z_2 < \cdots < z_n$. As in Section 1.4 we associate level k highest weights $\lambda_1, \cdots, \lambda_n$ to $p_1, \cdots, p_n$. It will often be convenient to take extra points $p_0 = O, p_{n+1} = \infty$ with highest weights $\lambda_0 = 0$ and $\lambda_{n+1} = 0$ and to identify the above space of conformal blocks with

$$\mathcal{H}(p_0, p_1, \cdots, p_{n+1}; \lambda_0, \lambda_1, \cdots, \lambda_{n+1}^*).$$

As shown in Lemma 1.17 the above space of conformal blocks is embedded in the space of coinvariant tensors $\mathrm{Hom}_{\mathfrak{g}}(\bigotimes_{j=1}^{n} V_{\lambda_j}, \mathbf{C})$. We identify the above space of conformal blocks with its image and we denote it simply by $V_{\lambda_1 \cdots \lambda_n}$. Let us recall that $V_{\lambda_1 \cdots \lambda_n}$ has a basis $\{v_{\mu_0 \mu_1 \cdots \mu_n}\}$ such that any triple $(\mu_{j-1}, \lambda_j, \mu_j)$ satisfies the quantum Clebsch-Gordan condition at level k.

For a generator σ_i of the braid group B_n the holonomy of the KZ connection along a path in $\mathrm{Conf}_n(\mathbf{C})$ representing σ_i defines a linear map

$$\rho(\sigma_i) : V_{\lambda_1 \cdots \lambda_i \lambda_{i+1} \cdots \lambda_n} \to V_{\lambda_1 \cdots \lambda_{i+1} \lambda_i \cdots \lambda_n}$$

where $V_{\lambda_1 \cdots \lambda_{i+1} \lambda_i \cdots \lambda_n}$ stands for the space of conformal blocks obtained by exchanging p_i and p_{i+1} as depicted in Figure 2.5. The above linear map depends only on the homotopy class of a path since the KZ connection is flat. In general, for $\sigma \in B_n$ the holonomy of the KZ connection along a path representing σ defines a linear map

$$\rho(\sigma) : V_{\lambda_1 \cdots \lambda_n} \to V_{\lambda_{\pi \circ \sigma(1)} \cdots \lambda_{\pi \circ \sigma(n)}}$$

where $\pi : B_n \to \mathcal{S}_n$ is a natural surjection. Considering $\rho(\sigma)$ as a right action, we have

$$\rho(\sigma \tau) = \rho(\sigma)\rho(\tau), \quad \sigma, \tau \in B_n.$$

A local monodromy is described by choosing a specific basis of the space of conformal blocks. We first deal with the case $n = 3$. For the solution Φ_1 of the KZ equation in (2.11) the action of the braid σ_1 is described as $P_{12} \exp(\pi \sqrt{-1} \Omega_{12}/\kappa)$ where P_{12} is the permutation operator

$$P_{12} : V_{\lambda_1 \lambda_2 \lambda_3} \to V_{\lambda_2 \lambda_1 \lambda_3}.$$

If we take a basis associated with the tree of type $(((12)3)4)$ with a labelling λ for the internal edge, then the matrix Ω_{12}/κ is diagonalized with eigenvalues $\Delta_\lambda - \Delta_{\lambda_1} - \Delta_{\lambda_2}$. On the other hand, the action of the braid σ_2 on the solution Φ_2 in (2.13) is written as

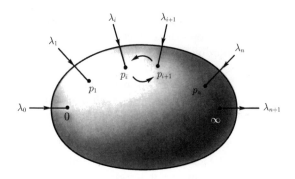

FIGURE 2.5. Action of the braid σ_i on the space of conformal blocks

$P_{23} \exp(\pi\sqrt{-1}\Omega_{23}/\kappa)$ where the matrix Ω_{23}/κ is diagonalized with respect to a basis associated with the tree of type $((1(23))4)$. To combine these local monodromies we need the connection matrix F satisfying $\Phi_1 = \Phi_2 F$. In general, for any n the computation of the monodromy representation of the braid group B_n on the space of conformal blocks can be reduced to the description of local monodromies and elementary connection matrices F.

DEFINITION 2.8. Let q be a non-zero complex number. The algebra over $\mathbf{C}$ with generators $1, T_i$, $1 \leq i \leq n-1$, and relations

$$T_i T_{i+1} T_i = T_{i+1} T_i T_{i+1}, \quad i = 1, \cdots, n-2,$$
$$T_i T_j = T_j T_i, \quad |i - j| > 1,$$
$$(T_i - q^{\frac{1}{2}})(T_i + q^{-\frac{1}{2}}) = 0$$

is called the *Iwahori-Hecke algebra* and is denoted by $H_n(q)$.

We put $q^{\frac{1}{2}} = \exp(\pi\sqrt{-1}/\kappa)$, $\eta = \exp(\pi\sqrt{-1}/2\kappa)$. The following theorem describes a relation between the monodromy representations of the braid groups on the space of conformal blocks and Iwahori-Hecke algebra representations.

THEOREM 2.9. *Let $\mathfrak{g}$ be the Lie algebra $sl_2(\mathbf{C})$ and $V_{\lambda_1}, \cdots, V_{\lambda_n}$ the standard 2-dimensional representations of $\mathfrak{g}$. Then the representation*

$$\tilde{\rho} : B_n \to GL(V_{\lambda_1 \cdots \lambda_n})$$

defined by $\tilde{\rho}(\sigma_i) = \eta\rho(\sigma_i), 1 \leq i \leq n-1$, *is an Iwahori-Hecke algebra representation in the sense that* $g_i = \tilde{\rho}(\sigma_i)$, $1 \leq i \leq n-1$, *satisfies*

$$(g_i - q^{\frac{1}{2}})(g_i + q^{-\frac{1}{2}}) = 0, \ 1 \leq i \leq n-1.$$

PROOF. Using the solution Φ_1 in (2.18), we see that the matrix $g_1 = \eta\rho(\sigma_1)$ is diagonalizable. Since the possible eigenvalues of Ω_{12}/κ are $\Delta_\lambda - 2\Delta_1$, $\lambda = 0, 2$, we obtain

$$(g_1 - q^{\frac{1}{2}})(g_1 + q^{-\frac{1}{2}}) = 0.$$

Similarly, using the solution Φ_2 associated with the tree of type $(\cdots(1(23))4)\cdots)$ we see that g_2 also satisfies $(g_2 - q^{\frac{1}{2}})(g_2 + q^{-\frac{1}{2}}) = 0$. The statement can be shown by applying the above argument for the other types of trees. $\square$

Our basis of the space of conformal blocks is in one-to-one correspondence with the sequence of integers

$$(\mu_0, \mu_1, \cdots, \mu_{n+1})$$

satisfying the conditions

$$\mu_0 = \mu_{n+1} = 0,$$
$$|\mu_i - \mu_{i+1}| = 1, \ 0 \leq i \leq n,$$
$$0 \leq \mu_i \leq k, \ 1 \leq i \leq n.$$

Figure 2.6 is the case $k = 3, n = 7$ and each of the above sequences of integers is represented as a shortest path connecting P and Q.

As we have seen, an explicit computation of the monodromy representation of the braid group can be reduced to the computation of the elementary connection matrix F for the equation (2.16) with $\lambda_2 = \lambda_3 = 1$. In this case the space of conformal blocks has dimension at most 2, and by computing the matrices Ω_{12} and Ω_{23} we can express the solutions by the Gauss hypergeometric functions. The connection matrix for the Gauss hypergeometric functions is well known. This allows us to compute the monodromy representation explicitly with respect to the above basis of the space of conformal blocks (see [51]). Theorem 2.9 can be generalized to the case $\mathfrak{g} = sl_m(\mathbf{C})$ and $V_{\lambda_1}, \cdots, V_{\lambda_n}$ are m-dimensional natural representations. In this case we have a similar result with $\kappa = k + m$ and $\eta = \exp(\pi\sqrt{-1}/m\kappa)$.

REMARK 2.10. As shown in Proposition 1.28, the space of conformal blocks has invariance under Möbius transformations. The space of conformal blocks $V_{\lambda_1\cdots\lambda_n}$ admits a linear action of the mapping class

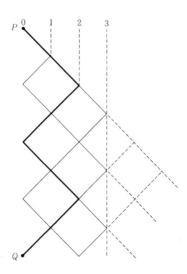

FIGURE 2.6. Basis of the space of conformal blocks

group $\mathcal{M}_{0,n}$ in addition to that of the braid group B_n. Here $\mathcal{M}_{0,n}$ stands for the mapping class group of a surface of genus 0 with n boundary components. The Dehn twists along curves C_j, $1 \leq j \leq n$, parallel to the boundary acts on $V_{\lambda_1 \cdots \lambda_n}$ by a scalar multiplication by $\exp(-2\pi\sqrt{-1}\Delta_j)$ (see [**35**]). We are going to deal with the action of the mapping class group on the space of conformal blocks in Section 2.4.

2.2. Conformal field theory and the Jones polynomial

In this section we describe a relation between the monodromy of the space of conformal blocks and the Jones polynomial. First, we briefly recall some basic materials for knots and links. A *link L* is an embedding

$$f : S^1 \sqcup \cdots \sqcup S^1 \to S^3$$

of a disjoint union of circles into S^3. The image of each S^1 is called a component of the link L. We write

$$L = L_1 \cup \cdots \cup L_m$$

where each L_j, $1 \leq j \leq m$, is a component of L. A link with one component is called a *knot*. Two links L and L' are called equivalent

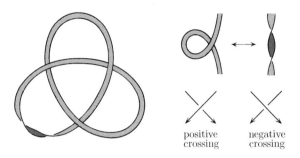

positive
crossing

negative
crossing

FIGURE 2.7. A trefoil knot with framing $+4$

if there exists an orientation preserving homeomorphism $h : S^3 \to S^3$ such that $L' = h(L)$.

Let $L = L_1 \cup L_2$ be a link with two components. We suppose that each component L_j, $j = 1, 2$, is oriented. The linking number $lk(L_1, L_2)$ is the intersection number of an oriented surface Σ_1 in S^3 such that $\partial\Sigma_1 = L_1$ and L_2. It is also computed from a projection diagram of L as half of the number of positive crossings of L_1 and L_2 minus the number of negative crossings of L_1 and L_2.

A *framing* of a link L is an integer n_j for each component L_j of L indicating a trivialization of the normal bundle of L_j. To specify the framing, we take a simple closed curve L'_j on the boundary of a tubular neighbourhood of L_j such that $lk(L_j, L'_j) = n_j$. The framing is indicated as in Figure 2.7 using a ribbon notation or a *blackboard framing*.

Let L be an oriented framed link in $\mathbf{R}^3$. To the components $L_1, \cdots, L_m$ of the link L one associates level k highest weights $\lambda_1, \cdots, \lambda_m$. We take a projection diagram of L in the region $x > 0$ of the xt plane. We suppose that the height function t on the link diagram has only non-degenerate critical points with distinct critical values. Furthermore, up to isotopy we may suppose that there exist horizontal lines $t = t_j$, $j = 0, 1, 2, \cdots, s$, such that each region $t_j < t < t_{j+1}$ contains one of the four types of diagrams: an elementary positive braid, an elementary negative braid, a minimal point or a maximal point, as shown in Figure 2.8. Such types of diagrams are called *elementary tangles*. Here we allow all possible orientations to the elementary tangles. We suppose that the intersection of the line $t = t_j$ for $j = 0, s$ and the link diagram is empty.

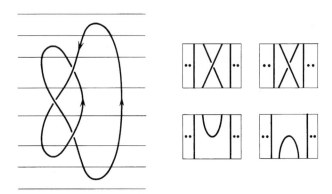

FIGURE 2.8. Decomposing a link diagram into elementary tangles

Let $q_1, \cdots, q_n$ be the intersection points of the link diagram with the line $t = t_j$, whose x coordinates $x_1, \cdots, x_n$ satisfy $0 < x_1 < \cdots < x_n$. To each q_i, $1 \leq i \leq n$, we associate a level k highest weight in the following way. If the component of the link L passing through q_i has a highest weight λ, then we associate to q_j the highest weight λ when the orientation of the link at q_j is downward. When the orientation is upward at q_j, we associate the dual weight λ^*. Then we consider the space of conformal blocks for the Riemann sphere with points $q_1, \cdots, q_n$ and the highest weights defined in the above way and we denote by $V(t_j)$ the space of conformal blocks at $t = t_j$ obtained in this manner. In particular, we have $V(t_0) = V(t_s) = \mathbf{C}$.

To each elementary tangle in Figure 2.8 we associate a linear map

$$(2.20) \qquad Z_j : V(t_j) \to V(t_{j+1}), \quad 0 \leq j \leq s - 1,$$

as follows. For an elementary tangle diagram representing the braid σ_i or σ_i^{-1}, we define Z_j to be the holonomy of the KZ equation along the corresponding braid. For maximal points and minimal points we add the extra weight 0 as in Figure 2.9 and we define Z_j by the elementary connection matrix F introduced in the last section in the following way. Consider the situation when the region

$$(x_i, x_{i+1}) \times (t_j, t_{j+1})$$

contains a minimal point as in Figure 2.9. We set

$$V(t_j) = V_{\lambda_1 \cdots \lambda_i \lambda_{i+1} \cdots \lambda_n}$$

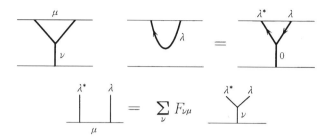

FIGURE 2.9. Representing minimal points by the connection matrix of the KZ equation

where each λ_j is λ or λ^*. Let us notice that $V(t_{j+1})$ is written as

$$V_{\lambda_1 \cdots \lambda_i \lambda \lambda^* \lambda_{i+1} \cdots \lambda_n}.$$

We denote by $\{v_{\mu_0 \cdots \mu_n}\}$ a basis of $V_{\lambda_1 \cdots \lambda_n}$ chosen as in Section 2.1. We have a natural identification

$$V_{\lambda_1 \cdots \lambda_i \lambda_{i+1} \cdots \lambda_n} \cong V_{\lambda_1 \cdots \lambda_i 0 \lambda_{i+1} \cdots \lambda_n}$$

by the map defined by

$$v_{\mu_0 \cdots \mu_i \cdots \mu_n} \mapsto v_{\mu_0 \cdots \mu_i \mu_i \cdots \mu_n}.$$

With this identification we define $Z_j : V(t_j) \to V(t_{j+1})$ by

$$v_{\mu_0 \cdots \mu_i \mu_i \cdots \mu_n} \mapsto \sum_\mu F_{\mu 0} \, v_{\mu_0 \cdots \mu_i \mu \mu_i \cdots \mu_n}$$

using the elementary fusing matrix as shown in Figure 2.9. The case of a maximal point is defined in a similar way.

Composing the above linear maps $Z_j, 0 \leq j \leq s - 1$, we put

(2.21) $Z(L; \lambda_1, \cdots, \lambda_m) = Z_{s-1} \circ \cdots \circ Z_1 \circ Z_0(1).$

The following lemma is a consequence of the flatness of the KZ connection.

LEMMA 2.11. *The above $Z(L; \lambda_1, \cdots, \lambda_m)$ is invariant under the local horizontal moves shown in Figure 2.10.*

Since $Z(L; \lambda_1, \cdots, \lambda_m)$ is not invariant under the operation of cancelling two critical points depicted in Figure 2.11, we need the following correction to obtain a link invariant. Let K_0 be a trivial

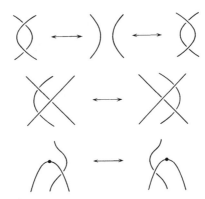

FIGURE 2.10. Local horizontal moves for links

FIGURE 2.11. Cancellation of two critical points

knot diagram with framing 0 which has two minimal points and two maximal points shown in Figure 2.11. We put

$$d(\lambda) = Z(K_0; \lambda)^{-1}.$$

Denoting by $\mu(j)$ the number of maximal points in the link component L_j, we set

$$(2.22) \quad J(L; \lambda_1, \cdots, \lambda_m) = d(\lambda_1)^{\mu(1)} \cdots d(\lambda_m)^{\mu(n)} Z(L; \lambda_1, \cdots, \lambda_m).$$

Let L be an oriented framed link. For L we fix a numbering of its components $L = L_1 \cup \cdots \cup L_m$. Such a link is called a colored oriented framed link. Two colored oriented framed links $L = L_1 \cup \cdots \cup L_m$ and $L' = L'_1 \cup \cdots \cup L'_m$ are called equivalent if there exists an orientation preserving homeomorphism h of S^3 such that

$$h(L_j) = L'_j, \ 1 \leq j \leq m,$$

where h preserves the orientation and the framing of each component.

THEOREM 2.12. $J(L; \lambda_1, \cdots, \lambda_m)$ is an invariant of a colored oriented framed link.

$$L_+ \qquad\qquad L_- \qquad\qquad L_0$$

FIGURE 2.12

PROOF. It is known that colored oriented framed links L and L' are equivalent if and only if L' is obtained from L by a sequence of horizontal local moves in Figure 2.10 and a cancellation of two critical points in Figure 2.11. By Lemma 2.11 it is enough to show the invariance under a cancellation of two critical points. Let L' be a link obtained from L by creating two critical points in the component L_j by the procedure shown in Figure 2.11. Then L' is represented as a connected sum of L and K_0. Hence we obtain

$$Z(L'; \lambda_1, \cdots, \lambda_m) = Z(K_0; \lambda_j) Z(L; \lambda_1, \cdots, \lambda_m).$$

It follows that $J(L; \lambda_1, \cdots, \lambda_m)$ is invariant under a cancellation of critical points. This completes the proof.   □

Let us recall that we showed in Proposition 1.28 the invariance of the horizontal sections of the KZ connection under Möbius transformations. In particular, the invariance under the dilatation $f(z) = \alpha z$ leads us to the following proposition.

PROPOSITION 2.13. *Let L' be a link obtained from $L = \bigcup_{j=1}^m L_j$ by increasing the framing of the component L_j by 1. Then,*

$$J(L'; \lambda_1, \cdots, \lambda_m) = \exp 2\pi\sqrt{-1}\Delta_{\lambda_j} J(L; \lambda_1, \cdots, \lambda_m)$$

holds.

In the case $\lambda_1 = \lambda_2 = \cdots = \lambda_m = \lambda$, $J(L, \lambda, \cdots, \lambda)$ is an invariant of an oriented framed link. In particular, when $\lambda_1 = \lambda_2 = \cdots = \lambda_m = 1$, we write J_L for $J(L, \lambda, \cdots, \lambda)$.

Let L_+, L_-, L_0 be oriented framed links which look locally in a 3-ball as in Figure 2.12 and are identical outside the 3-ball. The link invariants J_{L_+}, J_{L_-} and J_{L_0} satisfy a so-called *skein relation*. The skein relation for J_L is described as follows. Here we put

(2.23) $$q^{1/m} = \exp \frac{2\pi\sqrt{-1}}{m(k+2)}.$$

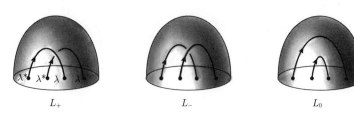

L_+ $\qquad\qquad\qquad$ L_- $\qquad\qquad\qquad$ L_0

FIGURE 2.13. The space of conformal blocks and the skein relation

PROPOSITION 2.14. *The link invariant J_L satisfies the skein relation*

$$q^{1/4} J_{L_+} - q^{-1/4} J_{L_-} = \left(q^{1/2} - q^{-1/2} \right) J_{L_0}.$$

PROOF. The monodromy matrix $G_i = \rho(\sigma_i)$ for the braid σ_i satisfies the quadratic relation

$$(G_i + q^{-3/4})(G_i - q^{1/4}) = 0$$

as shown in the proof of Theorem 2.9. It follows that

$$q^{1/4} G_i - q^{-1/4} G_i^{-1} = \left(q^{1/2} - q^{-1/2} \right) \, \mathrm{id}.$$

By definition the invariant J_L is obtained as the composition of linear maps associated with elementary tangles. We have decompositions into elementary tangles for the links L_+, L_- and L_0 such that they contain σ_i, σ_i^{-1} and a trivial braid respectively at the same level and are identical at the other parts. Therefore, the above relation for G_i implies the skein relation for J_L. This completes the proof. $\qquad\square$

The above skein relation is interpreted from the point of view of the space of conformal blocks in the following way. We localize the situation as in Figure 2.13 and consider the space of conformal blocks $\mathcal{H}$ of the Riemann sphere with four points and highest weights $\lambda_1, \lambda_2, \lambda_3, \lambda_4$. In our case these weights are given by $\lambda, \lambda, \lambda^*, \lambda^*$ where λ is the highest weight of the standard two dimensional representation of $sl_2(\mathbf{C})$. The diagrams L_+, L_- and L_0 determine vectors v_+, v_- and v_0 in the space of conformal blocks $\mathcal{H}$ respectively. It is clear from the construction that there is a vector w in the dual space $\mathcal{H}^*$ so that J_{L_+}, J_{L_-} and J_{L_0} are written as

$$J_{L_+} = \langle w, v_+ \rangle, \; J_{L_-} = \langle w, v_- \rangle, \; J_{L_0} = \langle w, v_0 \rangle$$

where $\langle\ ,\ \rangle : \mathcal{H}^* \times \mathcal{H} \to \mathbf{C}$ is the canonical pairing with the dual space. Since the dimension of the space of conformal blocks $\mathcal{H}$ is at most 2, we have a linear relation

$$\alpha v_+ + \beta v_- + \gamma v_0 = 0$$

with some $\alpha, \beta, \gamma \in \mathbf{C}$, which gives the relation

$$\alpha J_{L_+} + \beta J_{L_-} + \gamma J_{L_0} = 0.$$

The skein relation is nothing but an explicit form of the above linear relation.

The invariant J_L constructed above depends on the framing of a link. In order to obtain an invariant independent of the framing, we need the following correction. We start from a link diagram with a blackboard framing. We denote by $w(L)$ the number of positive crossings minus the number of negative crossings in the diagram. The number $w(L)$ is called the *writhe* of the framed link L. We set

$$P_L = d(1)^{-1} \exp\left(-2\pi\sqrt{-1}\Delta_1 w(L)\right) J_L.$$

It follows from Proposition 2.13 that P_L is an invariant of an oriented link which does not depend on the framing. Comparing the writhes $w(L_+)$, $w(L_-)$ and $w(L_0)$, we obtain from Proposition 2.14 the skein relation

$$q P_{L_+} - q^{-1} P_{L_-} = \left(q^{1/2} - q^{-1/2}\right) P_{L_0}.$$

The invariant P_L is characterized by the above skein relation and the value for a trivial knot

$$P_{\bigcirc} = 1.$$

This is a version of the Jones polynomial. The original *Jones polynomial* takes values in the ring of Laurent polynomials $\mathbf{Z}[t^{1/2}, t^{-1/2}]$ and is characterized by the skein relation and the value for a trivial knot

$$t^{-1}V_{L_+} - tV_{L_-} = \left(t^{1/2} - t^{-1/2}\right) V_{L_0},$$

$$V_{\bigcirc} = 1.$$

The invariant P_L is obtained from the Jones polynomial V_L by the substitution $t^{1/2} = -q^{-1/2}$. These special values of the Jones polynomial at roots of unity play an important role in the construction of Witten's invariants for 3-manifolds treated in the next section. To obtain the Jones polynomial with t indeterminate from conformal field

$$L_+ \qquad\qquad L_- \qquad\qquad L_0$$

FIGURE 2.14. Computation of $d(1)$ by the skein relation

theory we consider the case when the parameter k is generic. We replace the integrable highest weight module H_λ by the Verma module M_λ and perform a similar construction of link invariants.

Let us compute the normalization constant $d(\lambda)$ used in the above construction in the case $\lambda = 1$. For the links L_+, L_- and L_0 shown in Figure 2.14 we have $P_{L_+} = 1$, $P_{L_-} = 1$ and $P_{L_0} = d(1)$. Applying the skein relation, we obtain

$$d(1) = \frac{q - q^{-1}}{q^{1/2} - q^{-1/2}}.$$

So far, we have investigated the link invariant obtained by associating the highest weight of the standard 2-dimensional representation of $sl_2(\mathbf{C})$. In general, we color the components $L_1, \cdots, L_m$ of the link L with level k highest weights $\lambda_1, \cdots, \lambda_m$ and obtain an invariant of a colored oriented framed link $J(L; \lambda_1, \cdots, \lambda_m)$. The invariant $J(L; \lambda_1, \cdots, \lambda_m)$ will be computed from the invariant with highest weights 1 by the cabling operation and the fusion algebra R_k introduced in Section 1.6. Let K_0 be an oriented framed knot with a framing represented by an integer t. On the boundary of a tubular neighbourhood of K_0 we take an oriented knot K_1 in such a way that the linking number of K_0 and K_1 is equal to t. We give a framing t to the knot K_1. The above operation of adding a knot K_1 is called a *cabling*. In this way we obtain a two component link $K_0 \cup K_1$.

First, we compute the value $d(\lambda)$ for $\lambda > 1$.

LEMMA 2.15. *We have*

$$d(\lambda) = \frac{q^{(\lambda+1)/2} - q^{-(\lambda+1)/2}}{q^{1/2} - q^{-1/2}}$$

where $q^{1/2}$ is given in (2.23).

PROOF. By construction for a trivial knot K with 0-framing we have $Z(K; \lambda) = 1$, which implies $J(K; \lambda) = d(\lambda)$. It is clear that $d(0) = 1$. We have already shown the statement for $\lambda = 1$. Let us now compute $d(\lambda)$ for $\lambda > 1$. Considering the cabling for the trivial

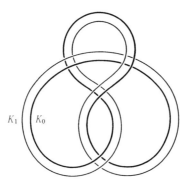

FIGURE 2.15. Cabling of the knot K_0 with a black-board framing

knot K with 0 framing, we have

$$(2.24) \qquad d(\lambda)d(\mu) = \sum_{\nu} N^{\nu}_{\lambda\mu} d(\nu).$$

Let us observe that

$$\frac{q^{(\lambda+1)/2} - q^{-(\lambda+1)/2}}{q^{1/2} - q^{-1/2}} = \frac{S_{0\lambda}}{S_{00}}$$

where $S_{\lambda\mu}$ is given in (1.44). We see that the values $d(\lambda), \lambda = 1, \cdots, k$, are characterized by the above equation (2.24) and the value $d(1)$. It will be enough to show that

$$d(\lambda) = \frac{S_{0\lambda}}{S_{00}}$$

actually satisfies the equation (2.24). This follows immediately from the Verlinde formula shown in Proposition 1.36. $\qquad \square$

Let K_0 be an oriented framed knot and let $K_0 \cup K_1$ be a link obtained by cabling of K_0. Generalizing the formula (2.24), we have the following result.

LEMMA 2.16. *The invariant* $J(K_0, K_1; \lambda, \mu)$ *of the link* $K_0 \cup K_1$ *obtained as a cabling of* K_0 *satisfies*

$$J(K_0, K_1; \lambda, \mu) = \sum_{\nu} N^{\nu}_{\lambda\mu} J(K; \nu)$$

where $N^{\nu}_{\lambda\mu}$ *is the structure constant of the fusion algebra* R_k.

For an oriented framed link L with m components one can define an invariant $J(L; x_1, \cdots, x_m)$ by associating $x_1, \cdots, x_m \in R_k$ to the components $L_1, \cdots, L_m$ in the following way. In the case $x_j = v_{\lambda_j}$, $j = 1, \cdots, m$, we define $J(L; x_1, \cdots, x_m)$ by $J(L; \lambda_1, \cdots, \lambda_m)$. When the component L_j is associated with the product $v_\lambda \cdot v_\mu$ we take the cabling of L_j and color these components with the weights λ and μ. By Lemma 2.16 we have

$$J(L; \cdots, v_\lambda \cdot v_\mu, \cdots) = \sum_\nu N_{\lambda\mu}^\nu J(K; \cdots, v_\nu, \cdots).$$

In this way we obtain a multilinear map

$$J(L) : R_k^{\otimes m} \to \mathbf{C}$$

defined by $J(L)(x_1, \cdots, x_m) = J(L; x_1, \cdots, x_m)$. In general, the computation of $J(L, \lambda_1, \cdots, \lambda_m)$ is reduced to that of invariants $J_{L'}$ for links L' obtained from L by the cabling operation.

In the following we sometimes simply write $J(L; \lambda)$ for

$$J(L; \lambda_1, \cdots, \lambda_m).$$

Let L_1 and L_2 be oriented framed links such that $L_1 \cap L_2 = \emptyset$. We consider the link $L_1 \cup L_2$ obtained as the union of L_1 and L_2. Let μ_1 and μ_2 denote colors for L_1 and L_2 respectively. We denote by $J(L_1 \cup L_2; \mu_1 \cup \mu_2)$ the invariant with the above colors.

PROPOSITION 2.17. *For links L_1 and L_2 contained in disjoint 3-balls B_1 and B_2 respectively*

$$J(L_1 \cup L_2; \mu_1 \cup \mu_2) = J(L_1; \mu_1)J(L_2; \mu_2)$$

holds.

PROOF. In the construction of $Z(L_1 \cup L_2; \mu_1 \cup \mu_2)$ we may put B_1 and B_2 in such a way that $Z(L_1 \cup L_2; \mu_1 \cup \mu_2)$ is equal to the composition $Z(L_1; \mu_1) \circ Z(L_2; \mu_2)$. After the correction coming from the number of maximal points we obtain the desired formula.   □

We denote by $\overline{L}$ the mirror image of L. Namely, $\overline{L}$ is obtained from L by applying an orientation reversing homeomorphism of S^3. The framing of $\overline{L}$ is equal to -1 times the framing of L.

PROPOSITION 2.18. *Let L be an oriented framed link. For the mirror image $\overline{L}$ of L we have*

$$J(\overline{L}; \lambda) = \overline{J(L; \lambda)}$$

where the right hand side stands for the complex conjugate of $J(L; \lambda)$.

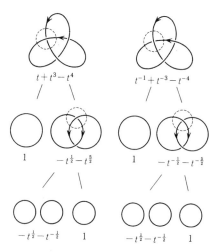

FIGURE 2.16. Jones polynomials for a right hand trefoil knot and its mirror image

PROOF. By the description of the monodromy representation in the last section the monodromy matrix $\rho(\sigma^{-1})$ is obtained from $\rho(\sigma)$ by replacing q with q^{-1}. The entries of the elementary connection matrix F and $d(\lambda)$ are real numbers. Therefore, we have $J_{\overline{L}}(q) = J_L(q^{-1})$. Since q is a root of unity this implies $J(\overline{L}; \lambda) = \overline{J(L; \lambda)}$. □

Applying the above argument in the case when q is a generic parameter we see that the Jones polynomial satisfies

$$V_{\overline{L}}(t) = V_L(t^{-1}).$$

We indicate in Figure 2.16 a computation of the Jones polynomials for a right hand trefoil knot and its mirror image using the skein relations.

The above construction for oriented framed links is generalized for oriented framed tangles. We recall the definition of tangles. Let X be the Cartesian product of the complex plane $\mathbf{C}$ and the unit interval $[0, 1]$. Let m and n be non-negative integers and take m distinct points $p_1, \cdots, p_m$ on the real line of $X_0 = \mathbf{C} \times \{0\}$ and n distinct points $q_1, \cdots, q_n$ on the real line of $X_1 = \mathbf{C} \times \{1\}$. A compact 1-manifold T in X with boundary $\{p_1, \cdots, p_m, q_1, \cdots, q_n\}$ is called an (m, n)-tangle. We represent a tangle by a projection diagram as in Figure 2.17. Each connected component of T is either a closed curve

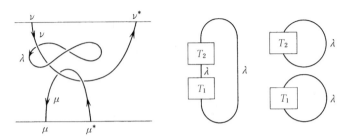

FIGURE 2.17. Tangle diagrams

or an arc whose end points are contained in $\{p_1, \cdots, p_m, q_1, \cdots, q_n\}$. We suppose that T is oriented and framed. An orientation is indicated by arrows and a framing is shown by a blackboard framing.

As in the case of a link diagram we can decompose a tangle diagram into elementary tangles up to isotopy equivalence. This allows us to define a linear map between the space of conformal blocks associated with a tangle in the following way. First, we color each arc of an (m, n)-tangle T with a level k highest weight. To each point $p_1, \cdots, p_m, q_1, \cdots, q_n$ we associate the highest weight for the arc passing through the point if the orientation of the arc is downward. If the orientation is upward, then we associate its dual to the point. We denote by $\mu_1, \cdots, \mu_m$ the level k highest weights for $p_1, \cdots, p_m$ obtained in this way. Similarly, the weights for $q_1, \cdots, q_n$ are given and we denote them by $\nu_1, \cdots, \nu_n$. Let $C_1, \cdots, C_l$ be the closed curves contained in the tangle. We color these curves with level k highest weights $\lambda_1, \cdots, \lambda_l$. By composing linear maps between the corresponding space of conformal blocks for elementary tangle diagrams, we obtain a linear map

$$Z(T; \lambda) : V_{\mu_1 \cdots \mu_m} \to V_{\nu_1 \cdots \nu_n}.$$

After a correction coming from the number of maximal points as in (2.22) we obtain a linear map

(2.25) $J(T; \lambda) : V_{\mu_1 \cdots \mu_m} \to V_{\nu_1 \cdots \nu_n}.$

This linear map is called the *tangle operator*. This tangle operator is considered to be a functor from the category of oriented framed tangles to the linear category. See [**31**] for the category of tangles. By using Lemma 2.15 and the construction of J we have the following proposition.

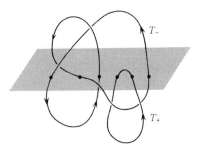

FIGURE 2.18. Decomposition of a link into two tangles T_+ and T_-

PROPOSITION 2.19. *Let T be an oriented framed $(1,1)$ tangle. We denote by $\widehat{T}$ a link obtained by closing T as in Figure 2.17. For the composition of tangles $T_1 \circ T_2$ with a color λ given as shown in Figure 2.17 we have*

$$J(\widehat{T_1 \circ T_2}) = J(\widehat{T_1})J(\widehat{T_2})\frac{S_{00}}{S_{0\lambda}}.$$

Let L be an oriented framed link with m components colored with highest weights $\lambda_1, \cdots, \lambda_m$. The link invariant $J(L,\lambda)$ may be interpreted in the following way by using tangle operators. We suppose up to isotopy that the link L intersects transversely with a horizontal plane H in such a way that the intersection points lie on a line. As in the construction of $J(L,\lambda)$ explained before, we associate the space of conformal blocks $V(t)$ to the above horizontal plane. By cutting the link into two parts by the plane H, we obtain two tangles T_+ and T_- as depicted in Figure 2.18. The image of 1 under the tangle operator associated with T_+ defines a vector in $V(t)$ denoted by $J(T_+)$. Similarly, looking at T_- upside down we obtain $J(L_-) \in V(t)^*$. Now the link invariant $J(L,\lambda)$ is expressed as

$$J(L;\lambda) = \langle J(L_+), J(L_-)\rangle$$

by means of the canonical paring between $V(t)$ and its dual space $V(t)^*$. This point of view will be treated later as an example of topological quantum field theory.

2.3. Witten's invariants for 3-manifolds

The purpose of this section is to formulate Witten's invariants for 3-manifolds based on conformal field theory. In the original seminal article [55], Witten introduced these invariants as partition functions of the Chern-Simons functionals.

Let L be a framed link in S^3. We first briefly recall a procedure called Dehn surgery to obtain a 3-manifold from L. We denote by $L_1 \cup \cdots \cup L_m$ the components of the link L. Let $N(L) = N(L_1) \cup \cdots \cup N(L_m)$ be a tubular neighbourhood of L. Each $N(L_j), 1 \le j \le m$, is homeomorphic to a solid torus $D \times S^1$. Take a closed curve γ_j on the boundary $\partial N(L_j)$ of $N(L_j)$ giving the framing of L_j. Namely, the linking number of L_j and γ_j is the integer corresponding to the framing of L_j. Take m copies of solid tori $H_j \cong D \times S^1$, $1 \le j \le m$, and a meridian m_j on the boundary of H_j. Here a meridian is a closed curve on the boundary of $D \times S^1$ expressed as $\partial D \times \{p\}$ with some $p \in S^1$ up to isotopy. We put $E(L) = \overline{S^3 \setminus N(L)}$, which is called the link exterior. The boundary of $E(L)$ is a disjoint union of tori

$$\partial E(L) = \partial N(L_1) \cup \cdots \cup \partial N(L_m).$$

Then, take a homeomorphism

$$f : \partial H_1 \cup \cdots \cup \partial H_m \to \partial N(L_1) \cup \cdots \cup \partial N(L_m)$$

such that $f(m_j) = \gamma_j, 1 \le j \le m$. We denote by M_L a 3-manifold obtained by attaching the link exterior $E(L)$ and the union of solid tori $H_1 \cup \cdots \cup H_m$ by the above homeomorphism f. The above procedure to obtain a 3-manifold M_L from a framed link L is called the *Dehn surgery* on L. The 3-manifold M_L has a natural orientation induced from that of S^3 and is determined uniquely up to orientation preserving homeomorphism by the equivalence class of a framed link L. It is known by Lickorish that any compact oriented closed 3-manifold is obtained by Dehn surgery on a framed link in S^3 (see [49]).

The above 3-manifold M_L is also expressed as the boundary of a 4-manifold in the following way. First we take a 4-ball B_4 with boundary S^3. We attach 2-handles to B_4 at $L \subset \partial B^4$ according to the framing of the link L and denote by W_L the obtained 4-manifold. Then, we have $M_L = \partial W_L$. Let us give some simple examples of Dehn surgeries. Dehn surgery on a trivial knot with framing 1 yields S^3. If we perform Dehn surgery on a trivial knot with framing 0, then we obtain $S^2 \times S^1$.

Let L and L' be framed links in S^3. Denote by M_L and M'_L the 3-manifolds obtained by Dehn surgery on L and L' respectively as above. The following is a formulation due to Fenn and Rourke [22] of Kirby's theorem.

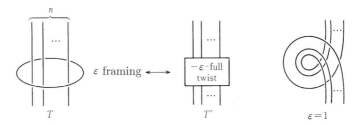

FIGURE 2.19. Kirby moves

THEOREM 2.20 (Kirby moves). *There is an orientation preserving homeomorphism $M_L \cong M'_L$ if and only if L' is obtained from L by applying local moves shown in Figure 2.19 finitely many times. Here, in Figure 2.19, $\varepsilon = \pm 1$ and n stands for the number of strands passing through the trivial knot with ε framing. In particular, in the case $n = 0$, this means a procedure of deleting or adding a trivial knot with ε framing contained in a 3-ball.*

Let L be an oriented framed link in S^3 with m components $L_1, \cdots, L_m$. For $1 \leq i \neq j \leq m$, we denote by $L_i \cdot L_j$ the linking number $lk(L_i, L_j)$. In the case $i = j$, $L_i \cdot L_i$ denotes the integer representing the framing of L_i. Let A be an m by m matrix with ij-entry $L_i \cdot L_j$. The matrix A is symmetric. Let n_+ (resp. n_-) be the number of positive (resp. negative) eigenvalues of A. The signature of the link L is defined to be $n_+ - n_-$ and is denoted by $\sigma(L)$. We observe that $\sigma(L)$ is equal to the signature of the intersection form of the 4-manifold W_L.

In order to define Witten's invariants for 3-manifolds we recall some notation from Chapter 1. Let $\mathfrak{g}$ be the Lie algebra $sl_2(\mathbf{C})$. Fix a positive integer k and denote by $P_+(k)$ the set of level k highest weights of the affine Lie algebra $\widehat{\mathfrak{g}}$. As in Chapter 1 we identify $P_+(k)$ with the set $\{0, 1, \cdots, k\}$. For each $\lambda \in P_+(k)$ one has an integrable highest weight module H_λ of $\widehat{\mathfrak{g}}$. The Virasoro Lie algebra acts on H_λ by Sugawara operators with central charge $c = \frac{3k}{k+2}$. Here we put

$$C = \exp\left(2\pi\sqrt{-1}\,\frac{c}{24}\right)^{-3} = \exp\left(-\pi\sqrt{-1}\,\frac{c}{4}\right).$$

The level k characters $\chi_\lambda(\tau)$, $\mathrm{Im}\,\tau > 0$, $\lambda \in P_+(k)$, of the affine Lie algebra $\widehat{\mathfrak{g}}$ behave under the action of $SL_2(\mathbf{Z})$ as

$$\chi_\lambda\left(-\frac{1}{\tau}\right) = \sum_\mu S_{\lambda\mu}\chi_\mu(\tau),$$

$$\chi_\lambda(\tau+1) = \exp 2\pi\sqrt{-1}\left(\Delta_\lambda - \frac{c}{24}\right)\chi_\lambda(\tau).$$

Here, as in (1.44) and (1.45), $S_{\lambda\mu}$ and Δ_λ are given explicitly by

$$S_{\lambda\mu} = \sqrt{\frac{2}{k+2}}\,\sin\frac{(\lambda+1)(\mu+1)}{k+2},$$

$$\Delta_\lambda = \frac{\lambda(\lambda+2)}{4(k+2)}.$$

These quantities play an essential role in the construction of Witten's invariants. The following lemma is an immediate consequence of the fact that the modular transformations S and T given by

$$S(\tau) = -\frac{1}{\tau}, \quad T(\tau) = \tau+1$$

satisfy $S^2 = (ST)^3 = I$.

LEMMA 2.21. *The above* $S_{\lambda\mu}$, $0 \le \lambda, \mu \le k$, *satisfy*

$$C\sum_\mu S_{\lambda\mu}S_{\mu\nu}\exp 2\pi\sqrt{-1}(\Delta_\lambda + \Delta_\mu + \Delta_\nu) = S_{\lambda\nu}.$$

Let L be an oriented framed link in S^3 with components $L_1, \cdots,$ L_m. Given a coloring $\lambda : \{1, \cdots, n\} \to P_+(k)$ with highest weights of level k, we defined in the last section the link invariant $J(L; \lambda_1, \cdots,$ $\lambda_m)$ where $\lambda_j = \lambda(j), 1 \le j \le m$. For a Hopf link the following holds.

PROPOSITION 2.22. *Let* H *be a Hopf link colored with* $\lambda, \mu \in$ $P_+(k)$ *as in Figure 2.20. Then, we have*

$$J(H; \lambda, \mu) = \frac{S_{\lambda\mu}}{S_{00}}.$$

PROOF. We represent a Hopf link as a cabling of a trivial knot with -1 framing shown in Figure 2.20. By Lemma 2.16 and Proposition 2.13 we have

$$\exp 2\pi\sqrt{-1}(-\Delta_\lambda - \Delta_\mu)J(H; \lambda, \mu) = \sum_\nu N_{\lambda\mu}^\nu \exp 2\pi\sqrt{-1}(-\Delta_\nu)\frac{S_{0\nu}}{S_{00}}.$$

Applying the Verlinde formula (Proposition 1.36) and Lemma 2.21 to the above formula, we obtain the desired result. $\qquad\square$

FIGURE 2.20. Hopf link

The following is a main result of this section.

THEOREM 2.23. *Let M be a compact oriented 3-manifold without boundary. Suppose that M is obtained as Dehn surgery on a framed link L with m components L_j, $1 \leq j \leq m$, in S^3. Then,*

$$Z_k(M) = S_{00}C^{\sigma(L)} \sum_{\lambda} S_{0\lambda_1} \cdots S_{0\lambda_m} J(L; \lambda_1, \cdots, \lambda_m)$$

is a topological invariant of M and does not depend on the choice of L which yields M. More precisely, if there is an orientation preserving homeomorphism $M_1 \cong M_2$, then $Z_k(M_1) = Z_k(M_2)$. Here, in the above formula, the sum is for any coloring $\lambda : \{1, \cdots, m\} \rightarrow P_+(k)$.

PROOF. Let us notice that in order to define the invariant $J(L; \lambda_1, \cdots, \lambda_m)$ we need an orientation for the link L, but the right hand side of the formula defining $Z_k(M)$ does not depend on the choice of an orientation.

We show that $Z_k(M)$ is invariant under Kirby moves. First, we deal with the case $n = 0$ in Figure 2.19. The case $\lambda = \nu = 0$ in Lemma 2.21 implies

$$C \sum_{\mu \in P_+(k)} S_{0\mu} \frac{S_{0\mu}}{S_{00}} \exp\left(2\pi\sqrt{-1}\Delta_\mu\right) = 1.$$

Combining with Lemma 2.15 and Proposition 2.13, we obtain the invariance of $Z_k(M)$ under the Kirby moves in the case $n = 0$.

To show the invariance under Kirby moves in the case $n = 1$ it will be enough to consider the case of a Hopf link by the factorization property in Proposition 2.19. By means of a result for a Hopf link in Proposition 2.22 and the formula in Lemma 2.21 for the case $\nu = 0$ written as

$$C \sum_{\mu \in P_+(k)} S_{0\mu} \frac{S_{\lambda\mu}}{S_{00}} \exp 2\pi\sqrt{-1}\Delta_\mu = \exp\left(-2\pi\sqrt{-1}\Delta_\lambda\right) \frac{S_{0\lambda}}{S_{00}}$$

we obtain the desired result.

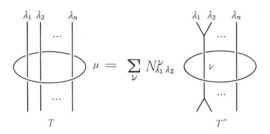

FIGURE 2.21

In the following, we show the invariance of $Z_k(M)$ under the Kirby moves by induction with respect to n. By using Proposition 2.19, we can localize the situation to the (n, n) tangles T and T' in Figure 2.19. We color each component of T and T' as in Figure 2.21. It will be sufficient to show the formula

$$C^{\sigma(L)} \sum_{\mu} S_{0\mu} J(T; \mu, \lambda_1, \cdots, \lambda_n) = C^{\sigma(L')} J(T'; \lambda_1, \cdots, \lambda_n).$$

Here L' denotes the link obtained from L by the above Kirby move. By using the method in Lemma 2.16, we can bind the two strands with coloring λ_1 and λ_2 as in Figure 2.21. Here, the trivalent vertices correspond to elementary fusing matrices. Thus, the tangle operator $J(T; \mu, \lambda_1, \cdots, \lambda_n)$ is expressed as a linear combination

$$\sum_{\nu} N^{\nu}_{\lambda_1 \lambda_2} F_1 J(T''; \mu, \nu, \lambda_3, \cdots, \lambda_n) F_2$$

where F_1 and F_2 are corresponding elementary fusing matrices and T'' is an $(n - 1, n - 1)$ tangle shown in Figure 2.21. Therefore, the desired equality for the tangle operators is reduced to the equality in the case of $n - 1$ strands. By the induction hypothesis and an investigation of the difference of $\sigma(L)$ under Kirby moves, we obtain the desired statement. This completes the proof. $\qquad \square$

The invariant $Z_k(M)$ thus obtained is called *Witten's invariant* for M associated with the Lie algebra $sl_2(\mathbf{C})$ at level k. Let us give some simple examples of $Z_k(M)$. In the case $M = S^3$ it follows immediately from the definition that

$$Z_k(S^3) = S_{00}.$$

For $M = S^1 \times S^2$ we make use of a presentation of $S^1 \times S^2$ as Dehn surgery on a trivial knot with 0 framing in S^3 and compute

$$Z_k(S^1 \times S^2) = S_{00} \sum_\mu S_{0\mu} \frac{S_{0\mu}}{S_{00}} = 1.$$

In the following, we give some basic properties of the invariant $Z_k(M)$.

PROPOSITION 2.24. *For a connected sum $M_1 \natural M_2$ of closed oriented 3-manifolds M_1 and M_2*

$$Z_k(M_1 \natural M_2) = \frac{1}{S_{00}} Z_k(M_1) Z_k(M_2)$$

holds.

PROOF. Let us suppose that M_1 and M_2 are obtained by Dehn surgery on framed links L_1 and L_2 respectively. Up to isotopy we take L_1 and L_2 in such a way that $L_1 \subset B_1$ and $L_2 \subset B_2$ where B_1 and B_2 are disjoint 3-balls in S^3. We see that the connected sum $M_1 \natural M_2$ is obtained as Dehn surgery on L. By using Proposition 2.17 and the definition of $Z_k(M)$ we obtain the formula.   □

PROPOSITION 2.25. *We denote by $-M$ the 3-manifold M with the orientation reversed. Then we have*

$$Z_k(-M) = \overline{Z_k(M)}.$$

PROOF. If M is obtained as Dehn surgery on a framed link L, then Dehn surgery on its mirror image $-L$ yields $-M$. Thus, by Proposition 2.18 we obtain the desired result.   □

Our method of constructing Witten's invariant $Z_k(M)$ is based on the monodromy of conformal field theory. The Dehn surgery formula for $Z_k(M)$ was first established by Reshetikhin and Turaev [48] by using representations of quantum groups at roots of unity.

We can extend the above construction to the case when a 3-manifold contains a link. Let M be a closed oriented manifold and L an oriented framed link in M. Let $L_1, \cdots, L_n$ be components of L with coloring $\lambda_1, \cdots, \lambda_n \in P_+(k)$. Suppose that (M, L) is obtained from (S^3, L') by Dehn surgery on a framed link N in S^3. Here L' is an oriented framed link in S^3 and we may assume up to isotopy that N does not intersect with L'. We denote by $N_1, \cdots, N_m$ the components of N. We define $Z_k(M, L; \lambda_1, \cdots, \lambda_n)$ by

$$Z_k(M, L; \lambda_1, \cdots, \lambda_n)$$
$$= S_{00} C^{\sigma(N)} \sum_\mu S_{0\mu_1} \cdots S_{0\mu_m} J(L' \cup N; \lambda_1, \cdots, \lambda_m, \mu_1, \cdots, \mu_m)$$

where the sum is for any coloring $\mu : \{1, \cdots, m\} \to P_+(k)$. It turns out that $Z_k(M, L; \lambda_1, \cdots, \lambda_n)$ is a topological invariant of the pair (M, L).

2.4. Projective representations of mapping class groups

The definition of Witten's invariant $Z_k(M)$ constructed in the last section is based on a Dehn surgery description of a 3-manifold. We will develop the construction in a wider setting, which will provide examples of *topological quantum field theories* in $2 + 1$ dimensions in the sense of Atiyah. Let us first list the axioms for topological quantum field theories. A topological quantum field theory in $d + 1$ dimensions is a functor Z satisfying the following conditions.

1. To each compact oriented d-dimensional smooth manifold without boundary Σ one associates a finite dimensional complex vector space Z_Σ.
2. A compact oriented $(d + 1)$-dimensional smooth manifold Y with $\partial Y = \Sigma$ determines a vector $Z(Y) \in Z_\Sigma$.

Furthermore, we suppose that Z satisfies the following properties (A1) – (A5).

(A1) We denote by $-\Sigma$ the manifold Σ with the orientation reversed. Then, we have $Z_{-\Sigma} = Z_\Sigma^*$ where Z_Σ^* is the dual of $Z_{-\Sigma}$ as a complex vector space.

(A2) For a disjoint union $\Sigma_1 \sqcup \Sigma_2$ we have $Z_{\Sigma_1 \sqcup \Sigma_2} = Z_{\Sigma_1} \otimes Z_{\Sigma_2}$.

It follows from the above (A1) and (A2) that a compact oriented $(d + 1)$-dimensional manifold Y with $\partial Y = (-\Sigma_1) \sqcup \Sigma_2$ determines a linear map $Z(Y) \in \mathrm{Hom}_{\mathbf{C}}(Z_{\Sigma_1}, Z_{\Sigma_2})$. Such a manifold Y is called a *cobordism* between Σ_1 and Σ_2.

(A3) For the composition of cobordisms $\partial Y_1 = (-\Sigma_1) \sqcup \Sigma_2$ and $\partial Y_2 = (-\Sigma_2) \sqcup \Sigma_3$

$$Z(Y_1 \cup Y_2) = Z(Y_2) \circ Z(Y_1)$$

holds where the right hand side stands for the composition of linear maps $Z(Y_1) : Z_{\Sigma_1} \to Z_{\Sigma_2}$ and $Z(Y_2) : Z_{\Sigma_2} \to Z_{\Sigma_3}$.

(A4) For an empty set $\emptyset$ we have $Z(\emptyset) = \mathbf{C}$.

(A5) Let I denote the closed unit interval. Then, $Z(\Sigma \times I)$ is the identity map as a linear transformation of Z_Σ.

Let $\mathcal{C}$ denote the category whose objects are compact oriented smooth manifolds without boundary and whose morphisms are cobordisms between such manifolds. The above Z is considered to be a functor from the category $\mathcal{C}$ to the linear category whose objects are finite dimensional complex vector spaces and whose morphisms are linear maps. Let us describe some of the direct consequences of the above axioms. We denote by $\mathrm{Diff}^+(\Sigma)$ the group of orientation preserving diffeomorphisms of Σ. From (A3) and (A5) it follows that $\mathrm{Diff}^+(\Sigma)$ acts linearly on Z_Σ. In particular, if Y is a closed manifold, then $Z(Y)$ is determined as a complex number. This number is also obtained in the following manner. Take a d-dimensional submanifold Σ and decompose Y into two parts Y_+ and Y_- by Σ. Namely, we take submanifolds with boundary Y_+ and Y_- such that $Y = Y_+ \cup Y_-$ and $\partial Y_+ = \Sigma = -\partial Y_-$. Then, we have vectors $Z(Y_+) \in Z_\Sigma, Z(Y_-) \in Z_\Sigma^*$ and $Z(Y)$ is given as

$$Z(Y) = \langle Z(Y_+), Z(Y_-) \rangle$$

by means of the canonical pairing between Z_Σ and Z_Σ^*. By (A3) this does not depend on the choice of a decomposition of Y. For $f \in \mathrm{Diff}^+(\Sigma)$ we denote by Σ_f the mapping torus obtained from $\Sigma \times I$ by identifying $x \in \Sigma \times \{0\}$ and $f(x) \in \Sigma \times \{1\}$. We have

$$Z(\Sigma_f) = \mathrm{Tr}\ \rho(f)$$

where $\rho(f)$ denotes the action of f on Z_Σ. In particular, considering the case when f is the identity map, we obtain

$$\dim Z_\Sigma = Z(\Sigma \times S^1).$$

The purpose of this section is to describe how the monodromy of the conformal field theory provides examples of topological quantum field theories in dimension $2 + 1$. In this case, as we will explain later, it is important to notice that the action of $\mathrm{Diff}^+(\Sigma)$ is not linear, but projectively linear.

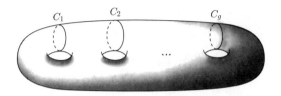

FIGURE 2.22. A cut system $C_1, \cdots, C_g$

Let Σ be a closed oriented surface of genus g. We introduce an equivalence relation $\sim$ on $\mathrm{Diff}^+(\Sigma)$ by defining $h \sim h'$ if and only if h and h' are isotopic. The set of equivalence classes $\mathrm{Diff}^+(\Sigma)/\sim$ forms a group by the composition of diffeomorphisms. This group is called the *mapping class group* of Σ and is denoted by $\mathcal{M}_g$. In Section 2.2, we have shown that the mapping class group $\mathcal{M}_{0,n}$ of a surface of genus 0 with n boundary components acts linearly on the space of conformal blocks of the Riemann sphere with n marked points. Here we generalize this construction to obtain representations of the mapping class group of a closed oriented surface of arbitrary genus.

For a closed oriented surface Σ of genus g we construct a complex vector space V_Σ in the following way. We regard Σ as the boundary of a handlebody H. Take disjoint simple closed curves $C_1, \cdots, C_g$ bounding discs in H as in Figure 2.22. We cut out the surface Σ by $C_1, \cdots, C_g$ to obtain a 2-sphere with $2g$ boundary components. Such a system of curves $C_1, \cdots, C_g$ is called a *cut system*. Let us now consider the space of conformal blocks of a Riemann sphere with $2g$ marked points. We take $2g$ points $p_1, p_2, \cdots, p_{2g-1}, p_{2g}$ on the real line of the complex plane so that their coordinates satisfy $0 < z_1 < z_2 < \cdots < z_{2g-1} < z_{2g}$. As in Section 2.2 we consider the case $\mathfrak{g} = sl_2(\mathbf{C})$ and fix a positive integer k. To $2g$ points $p_1, p_2, \cdots, p_{2g-1}, p_{2g}$ we associate level k highest weights $\mu_1, \mu_1^*, \cdots, \mu_g, \mu_g^*$. Here, as in Section 2.2, λ^* stands for the highest weight of the dual representation H_λ^* of H_λ. We associate H_0 to the origin and H_0^* to the point at infinity. As shown in Lemma 1.17, the corresponding space of conformal blocks is embedded in the space of coinvariant tensors

$$\mathrm{Hom}_\mathfrak{g}(V_{\mu_1} \otimes V_{\mu_1}^* \otimes \cdots \otimes V_{\mu_g} \otimes V_{\mu_g}^*, \mathbf{C}).$$

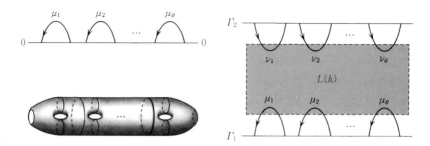

FIGURE 2.23. A trivalent graph Γ corresponding to a pants decomposition of Σ

We denote by $V_{\mu_1 \mu_1^* \cdots \mu_g \mu_g^*}$ the image of the space of conformal blocks by the above embedding and we set

$$V_\Sigma = \bigoplus_{\mu_1, \mu_1^*, \cdots, \mu_g, \mu_g^* \in P_+(k)} V_{\mu_1 \mu_1^* \cdots \mu_g \mu_g^*}.$$

A basis of V_Σ is in one-to-one correspondence with the set of admissible labellings for the edges of the graph Γ shown in Figure 2.23. The surface Σ is realized as the boundary of a regular neighbourhood the graph Γ embedded in $\mathbf{R}^3$. The graph Γ is considered to be the dual graph of a pants decomposition of Σ shown in Figure 2.23. The following dimension formula follows from Proposition 1.37.

LEMMA 2.26. *The above V_Σ is a finite dimensional complex vector space and we have*

$$\dim V_\Sigma = \sum_{\lambda \in P_+(k)} \left(\frac{1}{S_{0\lambda}} \right)^{2g-2}.$$

Let C be a simple closed curve on Σ and denote by τ_C the *Dehn twist* along C. Namely, τ_C is a diffeomorphism of Σ obtained by cutting Σ along C and by regluing after the 2π rotation as depicted in Figure 2.24. It is known by Lickorish (see [**14**]) that the mapping class group $\mathcal{M}_g$ is generated by the isotopy classes of Dehn twists along $3g - 1$ curves α_i, β_i, $1 \leq i \leq g$, δ_j, $2 \leq j \leq g$, in Figure 2.25.

By generalizing the definition of Witten's invariants of the last section to the case of 3-manifolds with boundary, we construct a projectively linear action of the mapping class group $\mathcal{M}_g$ on V_Σ. Let Γ_1 and Γ_2 be copies of the above graph Γ and embed Γ_1 and Γ_2 in $\mathbf{R}^2 \times I$ as shown in Figure 2.23. Let H_1 and H_2 be handlebodies obtained

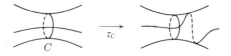

FIGURE 2.24. Dehn twist along C

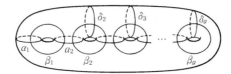

FIGURE 2.25. Lickorish generators

as regular neighbourhoods of Γ_1 and Γ_2 respectively. For $h \in \mathcal{M}_g$ we denote by M the 3-manifold obtained by gluing H_1 and H_2 by $h : \partial H_1 \to \partial H_2$. Consider $\mathbf{R}^2 \times I$ as a subset of $\mathbf{R}^3 \cup \{\infty\}$ and take a framed link L in $\mathbf{R}^2 \times I$ so that L does not intersect $\Gamma_1 \cup \Gamma_2$ and Dehn surgery on L yields M. This link L is denoted by $L(h)$. The graphs Γ_1 and Γ_2 together with the link $L(h)$ are considered to be a $(2g, 2g)$ framed tangle. The $(2g, 2g)$ tangles for Lickorish generators are shown in Figure 2.26. The composition of these generators is represented by the composition of the corresponding tangles. For $h \in \mathcal{M}_g$ express h as the composition of Lickorish generators and construct $L(h)$ by the composition of corresponding tangles for Lickorish generators.

Let $T(h)$ be the $(2g, 2g)$ tangle consisting of Γ_1, Γ_2 and $L(h)$. We choose an orientation of $L(h)$ and give a coloring by $\mu_1, \cdots, \mu_g$, $\nu_1, \cdots, \nu_g$ as shown in Figure 2.23. Given a coloring $\lambda : \{1, 2, \cdots, m\} \to P_+(k)$ for the components of $L(h)$, we obtain the tangle operator

$$J(T(h); \lambda)_{\mu\nu} : V_{\mu_1 \mu_1^* \cdots \mu_g \mu_g^*} \to V_{\nu_1 \nu_1^* \cdots \nu_g \nu_g^*}$$

as in Section 2.2. Using this tangle operator, we put

$$\rho(h)_{\mu\nu} = \sqrt{S_{0\mu_1} \cdots S_{0\mu_g}} \sqrt{S_{0\nu_1} \cdots S_{0\nu_g}}$$
$$\times C^{\sigma(L(h))} \sum_{\lambda} S_{0\lambda_1} \cdots S_{0\lambda_m} J(T(h); \lambda)_{\mu\nu}.$$

Here we use the notation of the last section. The above expression does not depend on the choice of an orientation for $L(h)$. We define

$$\rho(h) : V_\Sigma \to V_\Sigma$$

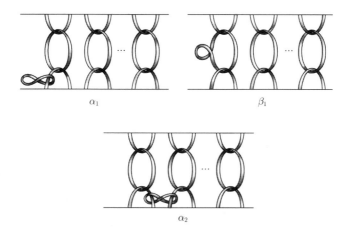

FIGURE 2.26. Tangles for Lickorish generators

by $\rho(h) = \bigoplus_{\mu\nu} \rho(h)_{\mu\nu}$. Thus, we obtain a map $\rho : \mathcal{M}_g \to GL(V_\Sigma)$. By using a version of Kirby moves for 3-manifolds with boundary, it can be shown as in Section 2.3 that $\rho(h)$ depends only on the isotopy class of h and not on the way of expressing h as the product of Lickorish generators.

For $x, y \in \mathcal{M}_g$ we consider the corresponding links $L(x)$, $L(y)$ and $L(xy)$ as above and we set

$$\xi(x, y) = C^{\sigma(xy) - \sigma(x) - \sigma(y)}.$$

Thus, we have shown the following proposition.

PROPOSITION 2.27. *The above map* $\rho : \mathcal{M}_g \to GL(V_\Sigma)$ *satisfies*

$$\rho(xy) = \xi(x, y)\rho(x)\rho(y)$$

for any $x, y \in \mathcal{M}_g$. *Namely,* ρ *is a projectively linear representation with the 2-cocycle* ξ.

Here ξ is a 2-cocycle of $\mathcal{M}_g$ with values in $\mathbf{C}^*$ in the sense that it satisfies

$$\xi(xy, z)\xi(x, y) = \xi(x, yz)\xi(y, z).$$

This follows immediately from $\rho((xy)z) = \rho(x(yz))$.

In the case $g = 1$, the mapping class group $\mathcal{M}_1$ is isomorphic to $SL_2(\mathbf{Z})$. We have $\dim V_\Sigma = k + 1$ and V_Σ has a basis $\{v_\lambda\}$ corresponding to the graph with one loop colored with λ, $0 \le \lambda \le k$.

LEMMA 2.28. *The action ρ of $SL_2(\mathbf{Z})$ on the above basis $\{v_\lambda\}$ is given by*

$$Sv_\lambda = \sum_\mu S_{\lambda\mu} v_\mu,$$

$$Tv_\lambda = \exp\left(2\pi\sqrt{-1}\Delta_\lambda\right) v_\lambda$$

where S and T are generators of $SL_2(\mathbf{Z})$ as in Section 2.3.

PROOF. We represent S and T respectively by $(2,2)$ tangles. Reducing the diagram for S by Kirby moves, we see that the result for S follows from the computation for the Hopf link in Proposition 2.22. The Dehn twist T is equal to α_1^{-1} and the formula follows from framing dependence in Proposition 2.13.   □

We observe that the action of $SL_2(\mathbf{Z})$ on the basis $\{v_\lambda\}$ coincides with the action on the Verlinde basis described in Section 1.6 up to a multiplication by some powers of C.

By successively applying elementary fusing operations in Section 2.1, we obtain bases of V_Σ corresponding to the admissible coloring of different trivalent graphs. These bases are related by connection matrices of the KZ equation. We take a regular neighbourhood of such a trivalent graph Γ' in $\mathbf{R}^3$, which is a handlebody H of genus g and the surface Σ is identified with its boundary. Let e be an edge of Γ' and let $D(e)$ be a 2-dimensional disc in H such that $\partial D(e) \subset \Sigma$ and $D(e)$ intersects transversely with e. It follows from the above construction of ρ that the Dehn twist along $C(e) = \partial D(e)$ acts diagonally on the basis of V_Σ corresponding to the admissible colorings of the edges of the graph Γ'. Let

$$\lambda : Edge(\Gamma') \to P_+(k)$$

be an admissible coloring where $Edge(\Gamma')$ is the set of edges of Γ', and let v_λ denote the corresponding vector in V_Σ. We have

$$\rho(\tau_{C(e)})v_\lambda = \exp\left(-2\pi\sqrt{-1}\Delta_{\lambda(e)}\right) v_\lambda.$$

For example, in the case of the graph Γ shown in Figure 2.23 the actions of $\alpha_1, \delta_2, \cdots, \delta_g$ are diagonalized simultaneously with respect to the basis corresponding to an admissible coloring of the edges of Γ.

Let M be a closed oriented 3-manifold. We now explain how Witten's invariant $Z_k(M)$ is recovered from the projective representation ρ. It is known that M is obtained from a handlebody H_g and its

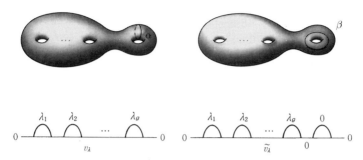

$$\text{FIGURE 2.27}$$

copy with the reversed orientation $-H_g$ by identifying their boundaries with a gluing map $h : \partial H_g \to \partial H_g$. Such a presentation of a closed oriented 3-manifold is called a *Heegaard splitting* and we write it by

$$M = H_g \cup_h (-H_g)$$

where h may be considered as an element of the mapping class group $\mathcal{M}_g$. The existence of a Heegaard splitting is shown by using a triangulation K of M and its dual triangulation K'. We take a regular neighbourhood H (resp. H') of the 1-skeleton of K (resp. K'). Both H and H' are handlebodies of the same genus (see [**49**]).

We denote by v_0 the vector corresponding to the coloring λ : $Edge(\Gamma) \to P_+(k)$ such that $\lambda(e) = 0$ for any $e \in Edge(\Gamma)$. We have the following description of $Z_k(M)$ based on a Heegaard splitting of M.

PROPOSITION 2.29. *Let* $M = H_g \cup_h (-H_g)$ *be a Heegaard splitting of a closed oriented 3-manifold* M. *Then, Witten's invariant* $Z_k(M)$ *is expressed as*

$$Z_k(M) = S_{00}^{-g+1} \langle v_0^*, \rho(h)v_0 \rangle$$

where v_0^* *is the dual element of* v_0 *in* V_Σ^* *and* $\langle \ , \ \rangle$ *is the canonical pairing between* V_Σ^* *and* V_Σ.

Let h be an element of the mapping class group $\mathcal{M}_g$. As in Figure 2.27 we add a 1-handle to the handlebody H_g to obtain H_{g+1}. We extend h to $\tilde{h} \in \mathcal{M}_{g+1}$ so that $\tilde{h}(\alpha) = \beta$ in Figure 2.27. The 3-manifold obtained as the Heegaard splitting

$$H_{g+1} \cup_{\tilde{h}} (-H_{g+1})$$

is a connected sum of $M = H_g \cup_h (-H_g)$ and S^3 and is homeomorphic to M. The above procedure of obtaining $H_{g+1} \cup_{\tilde{h}} (-H_{g+1})$ from $H_g \cup_h (-H_g)$ is called an elementary stabilization. Two Heegaard splittings $H_g \cup_h (-H_g)$ and $H_g \cup_{h'} (-H_g)$ are called equivalent if there exist $h_1, h_2 \in \mathcal{M}_g$ extended to homeomorphisms of the handlebody H_g such that

$$h' = h_1 \circ h \circ h_2.$$

Two Heegaard splittings are called *stably equivalent* if one is obtained from the other by a composition of elementary stabilizations or their inverse operations modulo the above equivalence. Stably equivalent Heegaard splittings give rise to homeomorphic 3-manifolds. If $H_g \cup_h (-H_g)$ and $H_{g'} \cup_{h'} (-H_{g'})$ are stably equivalent, then we can check directly that

$$S_{00}^{-g+1} \langle v_0^*, \rho(h)v_0 \rangle = S_{00}^{-g'+1} \langle v_0^*, \rho(h')v_0 \rangle$$

by using the equality

$$\langle v_0^*, \rho(\tilde{h})v_0 \rangle = S_{00} \langle v_0^*, \rho(h)v_0 \rangle$$

and the fact that an element of the mapping class group extended to a homeomorphism of the handlebody acts trivially on v_0 (see [**35**]).

Let M be a closed oriented 3-manifold. Considering a Morse function f on M, we decompose M into cobordisms between closed oriented surfaces. Let Σ be the inverse image of a regular value of f. We give an extra structure to Σ as the boundary of a handlebody and assign a complex vector space to Σ. When Σ is of genus g we define

$$V_\Sigma = \bigoplus_{\mu_1, \cdots, \mu_g \in P_+(k)} V_{\mu_1 \mu_1^* \cdots \mu_g \mu_g^*}.$$

We try to construct an example of topological quantum field theory in the sense of Atiyah. Here elementary cobordisms are gluing the boundary of handlebodies by an element of the mapping class group, attaching a handle, and its inverse operation. We described the action of the mapping class group in Proposition 2.27. For the procedure of attaching handles, the corresponding linear maps are defined in the following way. For attaching a 1-handle, we define $\iota : V_{\partial H_g} \to V_{\partial H_{g+1}}$ by

$$\iota(v_\lambda) = \frac{1}{\sqrt{S_{00}}} \tilde{v}_\lambda.$$

Here $\tilde{v}_\lambda$ is obtained from v_λ by adding a loop as in Figure 2.27 and labelling the loop by the weight 0. For the creation of S^2 we define $\beta : V_\emptyset \to V_{S^2}$ by

$$\beta(x) = \frac{x}{\sqrt{S_{00}}}.$$

In this way, one can associate a linear map to each cobordism. This gives an example of topological quantum field theory in $2+1$ dimensions except that the axiom (A3) holds up to a multiplication of a certain power of C.

Although we do not present details here, the above construction can be generalized to the case when a 3-manifold M contains a link L. We suppose that L is oriented and framed. To each component of L we associate a level k highest weight. Choose a generic Morse function f on M. Each level set of a regular value of f is a closed oriented surface Σ with finitely many marked points, say $p_1, \cdots, p_n$, corresponding to components of L as shown in Figure 2.28. Each point p_j, $1 \le j \le n$, is associated with a level k highest weight λ_j. We represent Σ as the boundary of a regular neighbourhood of the trivalent graph in Figure 2.28 and we set

$$(2.26) \qquad V_{\Sigma, p_1, \cdots, p_n, \lambda_1, \cdots, \lambda_n} = \bigoplus_{\mu_1, \cdots, \mu_g \in P_+(k)} V_{\mu_1 \mu_1^* \cdots \mu_g \mu_g^* \lambda_1 \cdots \lambda_n}.$$

We denote by $\mathcal{M}_{g,n}$ the mapping class group of an oriented surface of genus g with n boundary components. Generalizing Proposition 2.27, we can construct a projectively linear action of $\mathcal{M}_{g,n}$ on $V_{\Sigma, p_1, \cdots, p_n, \lambda_1, \cdots, \lambda_n}$. We can associate a linear map to each elementary cobordism and as a composition we obtain a topological invariant $Z(M, L; \lambda)$ for a colored oriented framed link L in M. In particular, in the case $M = S^3$, we have

$$(2.27) \qquad Z(M, L; \lambda) = S_{00} J(L, \lambda).$$

Consider the case of a solid torus H containing a knot K as its core with a highest weight λ. The vector space V_T assigned to the boundary $T = \partial H$ is $k+1$ dimensional and has a basis v_μ, $0 \le \mu \le k$. According to the above construction the solid torus H and the knot K with the highest weight λ determine a vector in V_T. We denote this vector by ϕ_λ. Since by Lemma 2.15

$$J(K, \lambda) = \frac{S_{0\lambda}}{S_{00}}$$

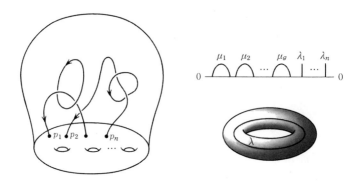

FIGURE 2.28

we obtain from (2.27) that

$$\langle S v_0, \phi_\lambda \rangle = S_{\lambda 0}.$$

Combining this with the action of $SL_2(\mathbf{Z})$ on V_T in Lemma 2.28, we conclude that $\phi_\lambda = v_\lambda$.

Let us discuss again the definition of Witten's invariant $Z_k(M)$ from this point of view. Let us suppose that M is obtained as Dehn surgery on a framed link L with m components in S^3. We denote by M_- the link exterior and by M_+ the union of m solid tori. We have $Z_{M_+} = Z_T^{\otimes m}$. For M_- we have a vector $Z_{M_-} \in (Z_T^{\otimes m})^*$ from the above construction of topological quantum field theory. Let f be the attaching map of Dehn surgery and write the action of f on $Z_T^{\otimes m}$ as

$$f v_{\mu_1} \otimes \cdots \otimes v_{\mu_m} = \sum_\lambda f_{\mu \lambda} v_{\lambda_1} \otimes \cdots \otimes v_{\lambda_m}.$$

Thus, up to a multiplication of a power of C,

$$\langle Z(M_+), Z(M_-) \rangle$$

is expressed as

$$\sum_\lambda f_{0\lambda} \langle Z_{M_-}, v_{\lambda_1} \otimes \cdots \otimes v_{\lambda_m} \rangle.$$

Furthermore, $\langle Z_{M_-}, v_{\lambda_1} \otimes \cdots \otimes v_{\lambda_m} \rangle$ is equal to $S_{00} J(L, \lambda)$ where L is given the 0-framing. In this way, we recover the definition of $Z_k(M)$ from topological quantum field theory up to a multiplication of a power of C.

2.5. Chern-Simons theory and connections on surfaces

Let M be a compact oriented 3-manifold, let G be the compact Lie group $SU(2)$ and let P be a principal G bundle over M. Since $SU(2)$ is simply connected, P is topologically a trivial $SU(2)$ bundle over M. We denote by $\mathcal{A}_M$ the space of connections on P. It turns out that $\mathcal{A}_M$ is an affine space, but we take a trivial connection as a base point and identify $\mathcal{A}_M$ with $\Omega^1(M, \mathfrak{g})$, the space of 1-forms on M with values in the Lie algebra $\mathfrak{g}$ of G. Denote by $\mathcal{G}$ the gauge group of P. It is identified with $\mathrm{Map}(M, G)$, the space of smooth maps from M to G. We have the right action of the gauge group $\mathcal{G}$ on the space of connections $\mathcal{A}_M$ defined by

$$g^* A = g^{-1} A g + g^{-1} dg, \quad A \in \mathcal{A}_M, g \in \mathcal{G}.$$

The curvature of a connection A is given by

$$F_A = dA + A \wedge A \in \Omega^2(M, \mathfrak{g}).$$

For $A \in \mathcal{A}_M$ we put

$$(2.28) \qquad CS(A) = \frac{1}{8\pi^2} \int_M \mathrm{Tr} \left(A \wedge dA + \frac{2}{3} A \wedge A \wedge A \right),$$

and we call the above CS the *Chern-Simons functional*. We refer the reader to [21] for details about classical Chern-Simons theory.

First, we suppose that the boundary of M is empty. We have the following proposition.

PROPOSITION 2.30. *A critical point of the Chern-Simons functional is a flat connection.*

PROOF. Consider a one-parameter family of connections $A_t = A + ta$. The Chern-Simons functional $CS(A + ta)$ is written in the form

$$(2.29) \qquad CS(A + ta) = CS(A) + \frac{t}{4\pi^2} \int_M \mathrm{Tr} \left(F_A \wedge a \right) + O(t^2)$$

where $F_A = dA + A \wedge A$ is the curvature of A. This shows that CS is critical at A if and only if $F_A = 0$. $\qquad \square$

The bilinear form $B : \Omega^1(M, \mathfrak{g}) \times \Omega^2(M, \mathfrak{g}) \to \mathbf{R}$ defined by

$$B(\alpha, \beta) = \int_M \mathrm{Tr}(\alpha \wedge \beta)$$

is non-degenerate and we consider $\Omega^2(M, \mathfrak{g})$ as the dual space of $\Omega^1(M, \mathfrak{g})$ by B. We denote by $F : \mathcal{A}_M \to \Omega^2(M, \mathfrak{g})$ the map given by $F(A) = F_A$. The tangent space of $\mathcal{A}_M$ is identified with $\Omega^1(M, \mathfrak{g})$ since $\mathcal{A}_M$ is an affine space. Combining with the above isomorphism $\Omega^2(M, \mathfrak{g}) \cong \Omega^1(M, \mathfrak{g})^*$, we see that F may be considered as a 1-form on $\mathcal{A}_M$. This 1-form F is exact since

$$(2.30) \qquad dCS(A) = \frac{1}{4\pi^2} F_A$$

holds by (2.29).

Let us now describe the action of the gauge group $\mathcal{G}$ on the Chern-Simons functional. For our purpose it will be necessary to consider the case $\partial M \neq \emptyset$.

PROPOSITION 2.31. *Let M be a compact oriented 3-manifold with $\partial M \neq \emptyset$. Then we have*

$$CS(g^*A) = CS(A) + \frac{1}{8\pi^2} \int_{\partial M} \mathrm{Tr}(A \wedge dg\ g^{-1}) - \int_M g^*\sigma$$

where σ is the volume form of $SU(2)$ in (1.8).

In particular, if $\partial M = \emptyset$, then we have

$$(2.31) \qquad CS(g^*A) = CS(A) - \int_M g^*\sigma.$$

The integral $\int_M g^*\sigma$ is considered as the mapping degree of the map $g : M \to SU(2)$ and is an integer. Therefore, the Chern-Simons functional induces a map

$$CS : \mathcal{A}_M/\mathcal{G} \to \mathbf{R}/\mathbf{Z}.$$

In the case $\partial M \neq \emptyset$ we set $\partial M = \Sigma$ and denote by Q a principal G bundle over Σ. Since we deal with the case $G = SU(2)$, the bundle Q is topologically isomorphic to a trivial bundle $\Sigma \times SU(2)$. We denote by $\mathcal{A}_\Sigma$ the space of connections on Q. As in the case of 3-manifolds $\mathcal{A}_\Sigma$ is identified with $\Omega^1(\Sigma, \mathfrak{g})$ and for $\alpha \in \mathcal{A}_\Sigma$ the tangent space $T_\alpha\mathcal{A}_\Sigma$ is isomorphic to $\Omega^1(\Sigma, \mathfrak{g})$. On the tangent space $T_\alpha\mathcal{A}_\Sigma$ we have an anti-symmetric bilinear form ω defined by

$$(2.32) \qquad \omega(\alpha, \beta) = -\frac{1}{8\pi^2} \int_\Sigma \mathrm{Tr}(\alpha \wedge \beta), \quad \alpha, \beta \in \Omega^1(\Sigma, \mathfrak{g}),$$

which defines a non-degenerate 2-form ω on $\mathcal{A}_\Sigma$ with $d\omega = 0$. In this sense $\mathcal{A}_\Sigma$ is equipped with a structure of an infinite dimensional symplectic manifold. We denote by $\mathcal{G}_\Sigma$ the gauge group of Q, which is identified with the space of smooth maps $\mathrm{Map}(\Sigma, G)$. In the following, we briefly describe a relation between the geometry of the space of connections $\mathcal{A}_\Sigma$ and the Chern-Simons functional for a 3-manifold.

For $a \in \mathcal{A}_\Sigma$ and $g \in \mathcal{G}_\Sigma$ we denote by A an extension of a on M as a connection of P and by $\tilde{g} : M \to G$ an extension of g as a smooth map on M with values in G. We put

$$(2.33) \qquad c(a,g) = \exp 2\pi\sqrt{-1}\left(CS(\tilde{g}^*A) - CS(A)\right).$$

More explicitly, $c(a,g)$ is written as

$$c(a,g) = \exp 2\pi\sqrt{-1}\left(\int_\Sigma \frac{1}{8\pi^2}\operatorname{Tr}(g^{-1}ag \wedge g^{-1}dg) - \int_M g^*\sigma\right).$$

The second term is the Wess-Zumino term in Section 1.3. The right hand side of the above equation does not depend on the choice of an extension $\tilde{g}$ and is determined uniquely by a and g on Σ. This can be formulated as the following proposition.

PROPOSITION 2.32. *Let M be a compact oriented 3-manifold with boundary Σ. For a connection A of a principal G bundle P over M and a gauge transformation $g \in \operatorname{Map}(M, G)$ we have*

$$\exp\left(2\pi\sqrt{-1}CS(g^*A)\right) = c(a, g|_\Sigma)\exp\left(2\pi\sqrt{-1}CS(A)\right)$$

where $g|_\Sigma$ denotes the restriction of g on Σ.

Let a be an element of $\mathcal{A}_\Sigma$. We define $L_{\Sigma,a}$ as the set of maps $f : \operatorname{Map}(\Sigma, G) \to \mathbf{C}$ satisfying

$$f(e \cdot g) = c(a,g)f(e), \quad g \in \operatorname{Map}(\Sigma, G).$$

Then $L_{\Sigma,a}$ is a 1-dimensional complex vector space with a Hermitian inner product. Let A be a connection of a principal G bundle over M and a its restriction on $\Sigma = \partial M$. We see from Proposition 2.32 that it is natural to consider $\exp\left(2\pi\sqrt{-1}CS(A)\right)$ as an element of $L_{\Sigma,a}$. We denote by $-\Sigma$ the closed oriented surface Σ with the orientation reversed. Then we have

$$L_{-\Sigma,a} \cong \overline{L_{\Sigma,a}}.$$

Let M be a closed oriented 3-manifold and suppose that M is decomposed as $M = M_1 \cup M_2$ with $\partial M_1 = \Sigma$ and $\partial M_2 = -\Sigma$ where $\Sigma \subset M$ is an embedded closed oriented surface. Let A be a connection of a principal G bundle on M and denote by A_1 and A_2 its restriction on M_1 and M_2 respectively. We denote by α the restriction of A on Σ. Then we have

$$\exp\left(2\pi\sqrt{-1}CS_{M_1}(A_1)\right) \in L_{\Sigma,a}, \quad \exp\left(2\pi\sqrt{-1}CS_{M_2}(A_2)\right) \in \overline{L_{\Sigma,a}}.$$

Using the canonical pairing $L_{\Sigma,\alpha} \times L_{-\Sigma,\alpha} \to \mathbf{C}$, we have

$$\exp\left(2\pi\sqrt{-1}CS_M(A)\right)$$
$$=\langle\exp\left(2\pi\sqrt{-1}CS_{M_1}(A_1)\right), \exp\left(2\pi\sqrt{-1}CS_{M_2}(A_2)\right)\rangle.$$

Let us consider a one-parameter family of connections of a principal G bundle Q over Σ and denote it by η_t, $0 \leq t \leq 1$. We regard η_t, $0 \leq t \leq 1$, as a connection η over $\Sigma \times [0,1]$. It follows from the above argument that the Chern-Simons functional $CS_{\Sigma\times[0,1]}$ over $\Sigma \times [0,1]$ defines a map

$$\exp(2\pi\sqrt{-1}CS_{\Sigma\times[0,1]}) : L_{\eta_0} \to L_{\eta_1}.$$

Let L_Σ be a topologically trivial complex line bundle over $\mathcal{A}_\Sigma$. For a path η_t, $0 \leq t \leq 1$, in $\mathcal{A}_\Sigma$ the above construction defines its lift on the total space of the complex line bundle L_Σ. Thus we obtain a connection on L_Σ whose horizontal sections are given by the above lift. It can be verified that as the curvature of this connection we recover the symplectic form ω on $\mathcal{A}_\Sigma$.

In this way, by means of the Chern-Simons functional, we can lift a path in the space of connections $\mathcal{A}_\Sigma$ to a path in the total space of the complex line bundle L_Σ. Similarly, using the Chern-Simons functional, we can lift the action of the gauge group $\mathcal{G}_\Sigma$ to L_Σ. By means of the moment map μ of this action we consider the Marsden-Weinstein quotient

$$\mathcal{A}_\Sigma//\mathcal{G}_\Sigma = \mu^{-1}(0)/\mathcal{G}_\Sigma.$$

Thus, we obtain a complex line bundle $\mathcal{L}$ on the moduli space $\mathcal{M}_\Sigma$ of flat G connections over Σ. The above moduli space $\mathcal{M}_\Sigma$ is identified with the set of equivalence classes of representations of the fundamental group

$$\mathrm{Hom}(\pi_1(\Sigma), G)/G.$$

Let M be a closed oriented 3-manifold and k an integer. Witten's invariant for 3-manifolds is formally written as the Feynman integral

$$Z_k(M) = \int_{\mathcal{A}_M/\mathcal{G}} \exp\left(2\pi\sqrt{-1}kCS(A)\right) \mathcal{D}A.$$

Our motivation was to formulate the partition function for a 3-manifold with boundary. Let us now suppose that M is an oriented 3-manifold with boundary Σ. We take A, a G connection over M, and we denote by α the restriction of A on Σ. We have shown that

$$\exp\left(2\pi\sqrt{-1}kCS(A)\right)$$

is considered to be an element of $L_{\Sigma,\alpha}$. We denote by $\mathcal{A}_{M,\alpha}$ the space of G connections over M whose restriction on Σ coincides with α. Consider formally the above Feynman integral restricted on $\mathcal{A}_{M,\alpha}$. Since

$$\exp\left(2\pi\sqrt{-1}kCS(g^*A)\right) = c(g,\alpha)^k \exp\left(2\pi\sqrt{-1}kCS(A)\right)$$

the partition function $Z_k(M)$ is considered to be a section of the complex line bundle $\mathcal{L}^{\otimes k}$.

The moduli space of flat connections $\mathcal{M}_\Sigma$ has a complex structure in the following way. As shown by Atiyah and Bott [7] the moduli space $\mathcal{M}_\Sigma$ is identified with the moduli space of topologically trivial holomorphic $G_{\mathbf{C}}$ bundles over Σ. This moduli space has a complex structure induced from a complex structure J on Σ. We denote by $\mathcal{M}_\Sigma^{(J)}$ the moduli space $\mathcal{M}_\Sigma$ with the above complex structure. Furthermore, we introduce the Kähler polarization and we denote by $\mathcal{H}_\Sigma$ the space of holomorphic sections of $\mathcal{L}^{\otimes k}$. It is known that $\mathcal{H}_\Sigma$ is a finite dimensional complex vector space and is isomorphic to the space of conformal blocks for Σ. A full exposition of this fact is out of the scope of this book. We refer the reader to [12]. In this way we can find a relationship between Chern-Simons gauge theory and conformal field theory. In particular, as a complex vector space, $\mathcal{H}_\Sigma$ is isomorphic to Z_Σ defined in Section 2.4.

Let us briefly mention the case of a link L in a 3-manifold M. Let C_j, $1 \le j \le r$, be the components of L and assign a representation R_j of the Lie group G to each component C_j. We define the Wilson line operator on the space of connections $\mathcal{A}_M$ by

$$W_{C_j,R_j}(A) = \mathrm{Tr}_{R_j} \mathrm{Hol}_{C_j}(A),$$

which stands for the trace on R_j of the holonomy of the connection A along C_j. Witten's invariant is written as

$$Z_k(M; C_1, \cdots, C_r) = \int \exp\left(2\pi\sqrt{-1}kCS(A)\right) \prod_{j=1}^r W_{C_j,R_j}(A)\mathcal{D}A.$$

The quantum Hilbert space associated with the above integral is formulated by means of the moduli space of parabolic bundles.

CHAPTER 3

Chern-Simons Perturbation Theory

In this chapter we deal with topological invariants obtained from the perturbative expansion of the partition function of the Chern-Simons functional as the level k tends to infinity. As the coefficients of the expansion of the Jones polynomial we obtain a series of finite type invariants for links. These are examples of Vassiliev invariants. In Section 3.1, we formulate the notion of the Vassiliev invariants and give a universal integral representation of Vassiliev invariants by means of the iterated integral due to Kontsevich. In Section 3.2, we investigate the behaviour of the Chern-Simons functional around a critical point. We describe a relation between the Hessian of the Chern-Simons functional at a critical point and the Ray-Singer torsion. This will suggest a principal term of the asymptotic expansion of Witten's invariant $Z_k(M)$ as $k \to \infty$. In Section 3.3, we proceed to explain a method of asymptotic expansions based on Feynman diagrams in a finite dimensional case. Motivated by this finite dimensional analogy we discuss Chern-Simons perturbative invariants for 3-manifolds due to Axelrod and Singer.

3.1. Vassiliev invariants and the Kontsevich integral

Vassiliev invariants were discovered in the study of the cohomology of the space of knots (see [54]). Let $\mathcal{M}$ be the space of smooth maps from S^1 to S^3 and Σ the subset of $\mathcal{M}$ consisting of $f : S^1 \to S^3$ such that f is not an embedding. The set Σ is called the *discriminant set*. We fix an orientation for S^1. The complement $\mathcal{M} \setminus \Sigma$ consists of the space of oriented knots and each connected component of $\mathcal{M} \setminus \Sigma$ is identified with a knot type. The 0-dimensional cohomology $H^0(\mathcal{M} \setminus \Sigma, \mathbf{C})$ may be considered to be the space of topological invariants of oriented knots with values in $\mathbf{C}$.

Vassiliev's idea developed in [54] is to approximate the space of knots by affine spaces consisiting of polynomial maps and to reduce the computation of the cohomology of $\mathcal{M} \setminus \Sigma$ to the cohomology of Σ by means of the Alexander duality. The discriminant set Σ is considered to be the space of singular knots. It has a filtration coming from the singularities of singular knots. This filtration allows us to define the spectral sequence which converges to the cohomology of $\mathcal{M} \setminus \Sigma$. A full account of Vassiliev's work is out of the scope of this book and we refer the reader to the original article [**54**]. The E_1 term consists of so-called invariants of finite order, which are formulated combinatorially in the following way.

We denote by $\mathcal{S}_n$ the subset of Σ consisting of singular knots with exactly n transversal double points. Let K be a singular knot defined by $f \in \mathcal{S}_n$ and denote its transversal double points by $p_1, \cdots, p_n$. For $\epsilon_j = \pm 1, j = 1, \cdots, n$, denote by $K_{\epsilon_1 \cdots \epsilon_n}$ the knot obtained from K by replacing the double point p_j with a positive crossing or a negative crossing according as $\epsilon_j = 1$ or $\epsilon_j = -1$. Let v be a topological invariant of oriented knots with values in $\mathbf{C}$. The invariant v is extended to singular knots with transversal double points by the rule

$$(3.1) \qquad v(K) = \sum_{\epsilon_j = \pm 1, j = 1, \cdots, n} \epsilon_1 \cdots \epsilon_n \, v(K_{\epsilon_1 \cdots \epsilon_n}).$$

We shall say that v is a *Vassiliev invariant* of order m if and only if v vanishes on $\mathcal{S}_n$ for any n such that $n > m$.

Let V_m denote the vector space of all Vassiliev invariants of order m for knots. It is easy to see that $V_0 = \mathbf{C}$ and that $V_0 = V_1$. We have an increasing series of vector spaces

$$\mathbf{C} = V_0 = V_1 \subset V_2 \subset \cdots \subset V_m \subset V_{m+1} \subset \cdots .$$

The vector space V_m will be described by so-called chord diagrams. Let us take a set of $2m$ distinct points $C = \{p_1, \cdots, p_{2m}\}$ on an oriented circle. We divide the set C into m sets $C_1, \cdots, C_m$ such that each set C_j consists of two points for any j, $1 \leq j \leq m$. Namely, we have

$$C = C_1 \cup \cdots \cup C_m, \quad C_i \cap C_j = \emptyset \ \text{ if } i \neq j,$$

and $|C_j| = 2$, $1 \leq j \leq m$. We connect two points in each C_j by a dotted line called a chord as shown in Figure 3.1. Such a diagram is called a *chord diagram* with m chords. We identify chord diagrams up to orientation preserving homeomorphisms. Given a singular knot $f : S^1 \to S^3$ with m transversal double points we obtain a chord diagram Γ with m chords by connecting each pair of points p and q

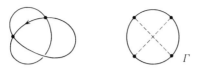

FIGURE 3.1. Chord diagram representation of a singular knot

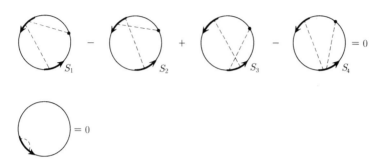

FIGURE 3.2. 4 term relation and framing independence relation

such that $f(p) = f(q)$ as depicted in Figure 3.1. We shall say that the singular knot defined by f respects the chord diagram Γ.

Let v be a Vassiliev invariant of order m. We take a chord diagram Γ with m chords and a singular knot K respecting Γ. Although the isotopy class of K is not determined by Γ, the value $v(K)$ defined by (3.1) is uniquely determined by Γ since v is of order m. Let $\mathcal{D}_m$ be a complex vector space whose basis is in one-to-one correspondence with the set of chord diagrams with m chords. By the above construction we obtain a linear map

$$w : V_m \to \mathrm{Hom}_{\mathbf{C}}(\mathcal{D}_m, \mathbf{C})$$

defined by $w(v)(\Gamma) = v(K)$ where K is a singular knot respecting Γ. We call $w(v)$ the *weight system* for the Vassiliev invariant v.

Let us now describe some important properties for weight systems. Let S_1, S_2, S_3, S_4 be four chord diagrams containing chords as shown in Figure 3.2. Here we suppose that there are no other chords with endpoints on the arcs with arrows in Figure 3.2. A chord connecting two adjacent points on a circle is called an isolated chord.

PROPOSITION 3.1. *Let v be a Vassiliev invariant of order m. Then we have the following properties.*

1. $\sum_{i=1}^{4}(-1)^{i-1}w(v)(S_i) = 0$.
2. *If S has an isolated chord, then $w(v)(S) = 0$.*

PROOF. We consider four singular knot diagrams respecting S_i, $1 \leq i \leq 4$. Each diagram contains two transversal double points corresponding to two chords in S_i. We resolve these double points by positive or negative crossings and apply (3.1) to express $w(v)(S_j)$ as an alternating sum of $w(v)$ for the resolved diagrams. The first relation follows immediately from this expression. The second equation follows from the fact that v is a topological invariant for oriented knots and does not depend on the framing. □

The first relation in Proposition 3.1 is called the *4 term relation* and the second one is called the *framing independence*. We denote by $\mathcal{A}_m$ the quotient space of $\mathcal{D}_m$ modulo the 4 term relations and the framing independence relations. By Proposition 3.1 the weight system defines an injective homomorphism

$$(3.2) \qquad V_m/V_{m-1} \to \mathrm{Hom}_{\mathbf{C}}(\mathcal{A}_m, \mathbf{C}).$$

In particular, it follows that V_m/V_{m-1} is a finite dimensional vector space. We have $\mathcal{A}_1 = 0$ since if a chord diagram has only one chord, it is an isolated chord. This confirms the assertion $V_0 = V_1$.

In [38] Kontsevich constructed an inverse to the map (3.2) using iterated integrals of logarithmic forms. It is a natural generalization of Chen's iterated integrals for braids. First, we recall the notion of iterated integrals of 1-forms. Let $\omega_1, \cdots, \omega_m$ be 1-forms on a smooth manifold M. For a path $\gamma : I \to M$ we denote by $\alpha_j(t)dt, 1 \leq j \leq m$, the pull back $\gamma^*\omega$. The iterated integral of $\omega_1, \cdots, \omega_m$ along the path γ is by definition the integral over the simplex

$$\int_{0 \leq t_m \leq \cdots \leq t_1 \leq 1} \alpha_1(t_1) \cdots \alpha_m(t_m)\, dt_1 \cdots dt_m$$

and is denoted by

$$\int_{\gamma} \omega_1 \cdots \omega_m.$$

We have

$$\int_{\gamma} \omega_1 \cdots \omega_m = (-1)^m \int_{\gamma} \omega_m \cdots \omega_1.$$

As in Section 2.1 we denote by $\mathrm{Conf}_n(\mathbf{C})$ the configuration space of n ordered distinct points in the complex plane $\mathbf{C}$. Consider the differential form

$$\omega_{ij} = d\log(z_i - z_j) = \frac{dz_i - dz_j}{z_i - z_j}, \quad i \neq j,$$

defined on $\mathrm{Conf}_n(\mathbf{C})$. Let κ be a non-zero complex parameter. We put

$$\omega = \frac{1}{\kappa} \sum_{1 \leq i < j \leq n} \Omega^{(ij)} \omega_{ij}$$

where $\Omega^{(ij)}$ is defined in Section 2.1. The monodromy of the KZ equation

$$(3.3) \qquad\qquad\qquad d\Phi = \omega\Phi$$

is expressed as an infinite sum of iterated integrals of ω by Picard's iteration in the following way. Let $\gamma(s), 0 \leq s \leq t$, be a path in $\mathrm{Conf}_n(\mathbf{C})$. Consider the solution of the total differential equation (3.3) along the path γ such that $\Phi(\gamma(0)) = I$. The equation (3.3) can be expressed as

$$\Phi(\gamma(t)) = I + \int_\gamma \omega\Phi(\gamma(s)).$$

We approximate the solution $\Phi(\gamma(t))$ by the sequence of functions $\Phi_n(\gamma(t))$, $n = 0, 1, \cdots$, defined inductively by

$$\Phi_{n+1}(\gamma(t)) = I + \int_\gamma \omega\Phi_n(\gamma(s)), \quad \Phi_0(\gamma(t)) = I.$$

Let $\gamma(t), 0 \leq t \leq 1$, be a loop in $\mathrm{Conf}_n(\mathbf{C})$. Then the holonomy of the connection ω along the loop γ is expressed as

$$(3.4) \qquad\qquad\qquad I + \sum_{m=1}^{\infty} \int_\gamma \underbrace{\omega \cdots \omega}_{m}.$$

As shown in Section 2.1 the flatness of the connection ω follows from the quadratic relations among $\Omega^{(ij)}$ in Lemma 2.1. Motivated by this, we introduce non-commutative indeterminates $X^{ij}, 1 \leq i \neq j \leq n$, in order to give a universal expression of the monodromy representation. We define A_n to be the algebra of non-commutative formal power series with indeterminates $X^{ij}, 1 \leq i \neq j \leq n$, over $\mathbf{C}$ modulo the two-sided ideal generated by $X^{ij} - X^{ji}$ and

(3.5) $[X^{ij} + X^{jk}, X^{ik}], \quad i, j, k$ distinct,

(3.6) $[X^{ij}, X^{kl}], \quad i, j, k, l$ distinct.

For a non-zero complex parameter κ consider a 1-form with values in A_n defined by

$$\tilde{\omega} = \frac{1}{\kappa} \sum_{1 \leq i < j \leq n} X^{ij} \omega_{ij}.$$

As in (3.4) we put

$$\tilde{\theta}(\gamma) = 1 + \sum_{m}^{\infty} \int_{\gamma} \underbrace{\tilde{\omega} \cdots \tilde{\omega}}_{m}.$$

Since $\tilde{\omega}$ is flat by (3.5) and (3.6), $\tilde{\theta}(\gamma)$ depends only on the homotopy class of the loop γ. Thus we have a homomorphism from the pure braid group P_n,

$$\tilde{\theta} : P_n \to A_n.$$

Here for $\gamma \in P_n$, $\tilde{\theta}(\gamma)$ is expressed as a formal power series in X^{ij}, whose coefficients are given by iterated integrals of logarithmic forms $\omega^{(ij)}$. The monodromy representation of the KZ equation is obtained by the substitution $X^{ij} = \Omega^{(ij)}$. In this sense $\tilde{\theta}$ is a universal expression of the monodromy representation of the KZ connection. Let us notice that the relation (3.5) is essentially the same as the 4 term relation.

Let $\gamma(t) = (z_1(t), \cdots, z_n(t))$ be a path in $\mathrm{Conf}_n(\mathbf{C})$ and denote the pull back $\gamma^* \omega_{ij}$ by

$$\omega_{ij}(t) = \frac{dz_i(t) - dz_j(t)}{z_i(t) - z_j(t)}.$$

The above $\tilde{\theta}$ is expressed explicitly as

$$\sum_{m=0}^{\infty} \frac{1}{\kappa^m} \int_{0 \leq t_m \leq \cdots \leq t_1 \leq 1} \sum_{P} X^{i_1 j_1} \cdots X^{i_m j_m} \omega_{i_1 j_1}(t_1) \wedge \cdots \wedge \omega_{i_k j_k}(t_k)$$

where the sum is for any $P = (i_1 j_1, \cdots, i_m j_m)$ such that $1 \leq i_1 < j_1 \leq n, \cdots, 1 \leq i_m < j_m \leq n$.

Generalizing the above iterated integrals for braids, Kontsevich obtained a universal integral expression for Vassiliev invariants of

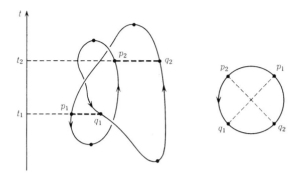

FIGURE 3.3

knots. Let K be an oriented knot in $\mathbf{R} \times \mathbf{C}$ and t a coordinate function for $\mathbf{R}$. We suppose that the critical points of the function t on K are non-degenerate. Namely, t is supposed to be a Morse function on K. We denote by $z_i(t)$ the curves connecting the maximal points and the minimal points of the height function t on K. In the case of Figure 3.3 we have $z_i(t)$, $i = 1, \cdots, 4$. We consider the iterated integral

$$Z_m(K) = \frac{1}{(2\pi\sqrt{-1})^m} \int_{t_1 < t_2 < \cdots < t_m} \sum_P (-1)^{\epsilon(P)} D_P \bigwedge_{k=1}^{m} \omega_{i_k j_k}(t_k).$$

Here the meaning of the notation is as follows. At each level $t = t_k$, $1 \leq k \leq m$, we choose a pair of points, say $\{z_{i_k}(t_k), z_{j_k}(t_k)\}$, continuously in t, and $P = (i_1 j_1, \cdots, i_m j_m)$ stands for such a choice of curves. The sum is for any possible choice P. Each P determines a chord diagram with m chords by connecting the above chosen points $\{z_{i_k}(t_k), z_{j_k}(t_k)\}$, $1 \leq k \leq m$, by dotted lines. We denote this chord diagram by D_P and we regard it as an element of $\mathcal{A}_m$. At the above $2m$ points $z_{i_k}(t_k), z_{j_k}(t_k), 1 \leq k \leq m$, we look at the orientation of the knot K, and denote by $\epsilon(P)$ the number of points where the orientation is downward.

The integral $Z_m(K)$ is considered to be an element of $\mathcal{A}_m$ and is called the *Kontsevich integral*. The convergence of the integral follows from the framing independence relation. Since the connection $\tilde{\omega}$ is integrable, we have the following lemma. See [11] for details.

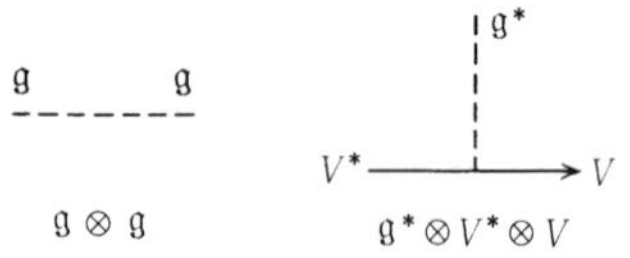

$$\mathfrak{g} \overset{\mathfrak{g}}{\text{-----}} \mathfrak{g}$$

$$\mathfrak{g} \otimes \mathfrak{g}$$

$$\begin{array}{c} \mathfrak{g}^* \\ \\ V^* \longrightarrow V \\ \\ \mathfrak{g}^* \otimes V^* \otimes V \end{array}$$

FIGURE 3.4. Lie algebra weight system

LEMMA 3.2. *The integral* $Z_m(K)$ *is convergent as an element of* $\mathcal{A}_m$ *and is invariant under local horizontal isotopy moves of the knot* K.

We denote by $\prod_{m=0}^{\infty} \mathcal{A}_m$ the direct product and we set

$$Z(K) = \sum_{m=0}^{\infty} Z_m(K) \in \prod_{m=0}^{\infty} \mathcal{A}_m.$$

The above infinite sum $Z(K)$ is not invariant under the cancellation of two critical points. In order to obtain a topological invariant for knots we make a correction as in Section 2.2. Let K_0 be a trivial knot with four critical points as depicted in Figure 2.11 in Section 2.2. We set

$$\tilde{Z}(K) = Z(K) \cdot Z(K_0)^{-\mu(k)+1}$$

where $\mu(K)$ is the number of maximal points in K. Let us notice that $Z(K_0) = \sum_{m=0}^{\infty} Z_m(K_0)$ is invertible in the direct product $\prod_{m=0}^{\infty} \mathcal{A}_m$. We express $\tilde{Z}(K)$ as

$$\tilde{Z}(K) = \sum_{m=0}^{\infty} \tilde{Z}_m(K), \quad \tilde{Z}_m(K) \in \mathcal{A}_m.$$

The following theorem is due to Kontsevich.

THEOREM 3.3 (Kontsevich). $\tilde{Z}(K)$ *is a topological invariant for oriented knots with values in* $\prod_{m=0}^{\infty} \mathcal{A}_m$. *Furthermore,* $\tilde{Z}_m(K)$ *is an invariant of order* m *with values in* $\mathcal{A}_m$.

Since the Kontsevich integral gives an inverse to the homomorphism (3.2) defined by the weight system, we have the following theorem.

THEOREM 3.4. *We have an isomorphism of vector spaces*

$$V_m/V_{m-1} \cong \mathrm{Hom}_{\mathbf{C}}(\mathcal{A}_m, \mathbf{C}).$$

REMARK 3.5. The notion of invariants of finite order can also be defined for braids. Let $V_m(P_n)$ be the space of order m invariants for the pure braid group P_n with values in $\mathbf{C}$. We denote by $\mathbf{C}[P_n]$ the group algebra over $\mathbf{C}$ of P_n. We have an isomorphism of vector spaces

$$V_m(P_n) \cong \operatorname{Hom}(\mathbf{C}[P_n]/I^{m+1}, \mathbf{C})$$

where I is the augmentation ideal of $\mathbf{C}[P_n]$. Taking the projective limit, we set

$$\widehat{\mathbf{C}[P_n]} = \varprojlim \mathbf{C}[P_n]/I^{k+1}.$$

Then we have an isomorphism of Hopf algebras

$$\widehat{\mathbf{C}[P_n]} \cong A_n.$$

We refer the reader to [**36**] for details.

An element of $\operatorname{Hom}(\mathcal{A}_m, \mathbf{C})$ is called a *weight system* of order m. It follows from Theorem 3.4 that for any weight system of order m there exists a corresponding Vassiliev invariant of order m. Given a complex semisimple Lie algebra $\mathfrak{g}$ and its finite dimensional representation $r : \mathfrak{g} \to \operatorname{End}(V)$ one can construct a weight system in the following way. We decompose a chord diagram into chords and trivalent vertices as in Figure 3.4. To each chord one associates

$$\frac{1}{\kappa}\Omega = \frac{1}{\kappa}\sum_{\mu} I_\mu \otimes I_\mu \in \mathfrak{g} \otimes \mathfrak{g}$$

where $\{I_\mu\}$ is an orthonormal basis of $\mathfrak{g}$ with respect to the Cartan-Killing form. To a trivalent vertex one associates an element in

$$\mathfrak{g}^* \otimes V^* \otimes V$$

corresponding to the representation r. Given a chord diagram Γ we contract the above tensors according to the diagram to obtain a scalar denoted by $\tau(\Gamma)$. One can verify that the above $\tau(\Gamma)$ satisfies the 4 term relation. The Lie algebra weight system for $\mathfrak{g}$ and r is obtained by normalizing τ so that it satisfies the framing independence relation. As an example consider the case when $\mathfrak{g} = sl_m(\mathbf{C})$ and r is the m-dimensional natural representation. As mentioned after Theorem 2.9, the monodromy representations of the braid group factor through the Iwahori-Hecke algebra, which leads to a skein relation to the link invariant associated with our weight system. It turns out that we obtain the two variable Jones polynomial $P(x, y)$ with $x = q^{m/2}$ and

$y = q^{-1/2} - q^{1/2}$. Let us recall that $P(x, y)$ is characterized by the skein relation

$$x^{-1}P_{L_+} - xP_{L_-} = yP_{L_0}, \quad P_{\bigcirc} = 1.$$

It follows that the normalization term $Z(K_0)$ is given by

$$(3.7) \qquad\qquad Z(K_0) = \frac{m(q^{1/2} - q^{-1/2})}{q^{m/2} - q^{-m/2}}.$$

The *Conway polynomial* $\nabla_L(z)$ is an invariant of oriented links characterized by the skein relation

$$\nabla_{L_+} - \nabla_{L_-} = z\nabla_{L_0}, \quad \nabla_{\bigcirc} = 1.$$

We see that $\nabla_L(z)$ is a polynomial in z. Let us extend the definition of the Vassiliev invariants to oriented links in a similar way as for the case of oriented knots. We denote by $a_k(L)$ the coefficient of z^k of the Conway polynomial $\nabla_L(z)$. It follows directly from the definition of the Vassiliev invariants and the above skein relation that a_k is a Vassiliev invariant of order k. Now we focus on the Vassiliev invariants of order 2 for oriented knots. The order 2 term $\tilde{Z}_2(K)$ of $\tilde{Z}(K)$ is written as

$$(3.8)$$
$$\frac{1}{4\pi^2} \int_{t_1 < t_2} \sum_P (-1)^{\epsilon(P)} D_P \omega_{i_1 j_1}(t_1) \wedge \omega_{i_2 j_2}(t_2) + \frac{1}{24}\mu(K) - \frac{1}{24}$$

where D_P is a chord diagram with two chords. We observe that $\dim \mathcal{A}_2 = 1$ and that $\mathcal{A}_2$ is spanned by the chord diagram D with two chords shown in Figure 3.3. Let w be an element of $\mathrm{Hom}_{\mathbf{C}}(\mathcal{A}_2, \mathbf{C})$ such that $w(D) = 1$ holds for the above chord diagram D. Applying w to the integral (3.8), we obtain a Vassiliev invariant of order 2 for oriented knots. Since the coefficient a_2 of the Conway polynomial is of order 2 and its weight system is above w, we conclude that $w(\tilde{Z}_2(K))$ coincides with $a_2(K)$. Thus we obtain an integral representation of $a_2(K)$.

To obtain the expression for $\tilde{Z}_2(K)$ in (3.8) we used the order 2 term of the Kontsevich integral $Z(K_0)$ for a trivial knot with four critical points K_0. As shown in Figure 3.5 this computation involves the iterated integral of the differential forms

$$\omega_0 = \frac{dt}{t}, \quad \omega_1 = \frac{dt}{t - 1}.$$

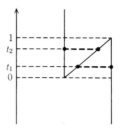

FIGURE 3.5. Order 2 term of $Z(K_0)$

The above iterated integral is expressed as a special value of the dilogarithm

$$\int_0^1 \frac{\log(1-t)}{t}\, dt = -\frac{\pi^2}{6}.$$

In general the higher terms of $Z(K_0)$ are expressed as iterated integrals of ω_0 and ω_1.

Let $k_1, \cdots, k_n$ be positive integers and suppose $k_n \geq 2$. We define the *multiple zeta value* $\zeta(k_1, \cdots, k_n)$ by

$$\zeta(k_1, \cdots, k_n) = \sum_{0 < m_1 < \cdots < m_n} \frac{1}{m_1^{k_1} \cdots m_n^{k_n}}.$$

We describe a relation between the multiple zeta values and the iterated integrals of ω_0 and ω_1. Let $k_1, \cdots, k_n$ be positive integers. We define the *polylogarithm* $L_{k_1 \cdots k_n}$ by

$$L_{k_1 \cdots k_n}(z) = \sum_{0 < m_1 < \cdots < m_n} \frac{z^{m_n}}{m_1^{k_1} \cdots m_n^{k_n}},$$

which is an analytic function in z in the domain $|z| < 1$. If $k_n \geq 2$, then $L_{k_1 \cdots k_n}(1)$ is convergent and coincides with the multiple zeta value $\zeta(k_1, k_2, \cdots, k_n)$. The polylogarithm $L_{k_1 \cdots k_n}(z)$ has the following property.

LEMMA 3.6. *If $k_n > 1$, then we have*

$$\frac{d}{dz} L_{k_1 \cdots k_n} = \frac{1}{z} L_{k_1 \cdots k_n - 1}.$$

In the case $k_n = 1$ we have

$$\frac{d}{dz} L_{k_1 \cdots k_n} = \frac{1}{1-z} L_{k_1 \cdots k_n - 1}.$$

PROOF. The case $k_n > 1$ is obvious. The formula for $k_n = 1$ follows from

$$\sum_{m_n=m_{n-1}+1}^{\infty} z^{m_n-1} = \frac{z^{m_{n-1}}}{1-z}.$$

$\square$

We start from

$$L_1(z) = \int_0^z \frac{dt}{1-t}$$

and repeatedly apply the above lemma to obtain the following proposition.

PROPOSITION 3.7. The polylogarithm $L_{k_1\cdots k_n}(z)$ is expressed by the integral

$$\int_0^z \underbrace{\omega_0 \cdots \omega_0}_{k_n-1} \omega_1 \underbrace{\omega_0 \cdots \omega_0}_{k_{n-1}-1} \omega_1 \cdots \omega_1 \underbrace{\omega_0 \cdots \omega_0}_{k_1-1} \omega_1.$$

In the following we write

$$\int_0^z \omega_0^{k_n-1} \omega_1 \omega_0^{k_{n-1}-1} \omega_1 \cdots \omega_1 \omega_0^{k_1-1} \omega_1$$

for the above iterated integral. It follows from Proposition 3.7 that

$$\zeta(k_1, k_2, \cdots, k_n) = \int_0^1 \omega_0^{k_n-1} \omega_1 \omega_0^{k_{n-1}-1} \omega_1 \cdots \omega_1 \omega_0^{k_1-1} \omega_1.$$

For $\varepsilon_j = 0$ or 1, $1 \leq j \leq k$, we denote by $I_{\varepsilon_1\cdots\varepsilon_k}$ the iterated integral

$$\int_0^1 \omega_{\varepsilon_1} \cdots \omega_{\varepsilon_k}.$$

By the change of variable

$$(t_1, \cdots, t_k) \mapsto (1 - t_1, \cdots, 1 - t_k)$$

we have

$$I_{\varepsilon_1\cdots\varepsilon_k} = I_{1-\varepsilon_k\cdots 1-\varepsilon_1},$$

which gives rise to so-called duality relations among mutiple zeta values. For example, we recover the relation $\zeta(1,2) = \zeta(3)$ due to Euler.

We describe some additional properties of the Drinfel'd associator introduced briefly in Section 2.1. First, let us give a precise definition. We denote by $\mathbf{C} \ll X, Y \gg$ the ring of non-commutative formal power series with indeterminates X and Y. Consider the differential equation

$$(3.9) \qquad G'(x) = \left(\frac{X}{x} + \frac{Y}{x-1} \right) G(x)$$

on the Riemann sphere. The differential equation (3.9) has regular singularities at 0, 1 and ∞. In the real interval $0 < x < 1$, we consider the solutions $G_1(x)$ and $G_2(x)$ of (3.9) with asymptotics

$$G_1(x) \sim x^X, \quad x \to 0,$$
$$G_2(x) \sim (1-x)^Y, \quad x \to 1,$$

where x^X stands for $\exp(X \log x)$. More precisely, $G_1(x)$ is a solution of the form

$$G_1(x) = P(x)x^X$$

where $P(x)$ is a power series

$$P(x) = 1 + \sum_{r>0} P_r x^r, \ P_r \in \mathbf{C} \ll X, Y \gg .$$

One can show that such a solution of (3.9) exists uniquely by determining P_r, $r > 0$, inductively. It follows from the general theory of differential equations in complex domains that $P(x)$ is convergent and is analytic in x. Similarly, $G_2(x)$ is a unique solution of the form

$$G_2(x) = Q(1-x)(1-x)^Y$$

where $Q(x)$ is an analytic function with $Q(1) = 1$. Since both $G_1(x)$ and $G_2(x)$ are solutions of the same differential equation (3.9), by analytic continuation there exists $\Phi(X,Y) \in \mathbf{C} \ll X, Y \gg$ such that

$$G_1(x) = G_2(x)\Phi(X,Y)$$

holds. Let us notice that $\Phi(X,Y)$ does not depend on x. We call $\Phi(X,Y)$ the Drinfel'd associator.

For a with $0 < a < 1$ we denote by $G^a(x)$ the unique solution of (3.9) such that $G^a(a) = 1$. The Drinfel'd associator $\Phi(X,Y)$ is expressed in terms of G^a in the following way.

LEMMA 3.8. *We have*

$$\Phi(X,Y) = \lim_{a \to 0} a^{-Y} G^a(1-a)a^X.$$

PROOF. The solution $G^a(x)$ is written as

$$G^a(x) = G_1(x)G_1(a)^{-1} = G_1(x)a^{-X}P(a)^{-1}.$$

Hence we have

$$a^{-Y}G^a(1-a)a^X = a^{-Y}G_1(1-a)a^{-X}P(a)^{-1}a^X$$
$$= a^{-Y}Q(a)a^Y\Phi(X,Y)a^{-X}P(a)^{-1}a^X.$$

The equation in our lemma follows immediately.   □

For non-negative integers $p_1, q_1, \cdots, p_k, q_k$ we set

$$I_a(p_1, q_1, \cdots, p_k, q_k) = \int_a^{1-a} \omega_0^{p_1}\omega_1^{q_1}\cdots\omega_0^{p_k}\omega_1^{q_k}.$$

We see that $G_a(1-a)$ is expressed by means of the above iterated integral as

$$G_a(1-a) = 1 + \sum I_a(p_1, q_1, \cdots, p_k, q_k)X^{p_1}Y^{q_1}\cdots X^{p_k}Y^{q_k}$$

where the sum is for any non-negative integers $p_1, q_1, \cdots, p_k, q_k$ with $k \geq 1$ such that at least one of them is non-zero. Let us notice that the integral $I_a(p_1, q_1, \cdots, p_k, q_k)$ is divergent as $a \to 0$ if and only if the corresponding monomial $X^{p_1}Y^{q_1}\cdots X^{p_k}Y^{q_k}$ starts with Y or ends with X. In order to deal with such divergent integrals we will make use of the following construction.

We put $\Lambda = \mathbf{C} \ll X, Y \gg$ and define Λ' as the submodule over $\mathbf{C}$ of Λ spanned by all monomials starting with X and ending with Y. We define $\pi : \Lambda \to \Lambda'$ to be the projection map. Consider the polynomial algebra $\Lambda[A, B]$ over Λ generated by commuting indeterminates A and B. Here we suppose that A and B commute with any element of Λ. Any monomial in $\Lambda[A, B]$ can be written uniquely as $B^q M A^p$ with a monomial $M \in \Lambda$ where p and q are non-negative integers. We define a $\mathbf{C}$-linear map $j : \Lambda[A, B] \to \Lambda$ by $j(B^q M A^p) = Y^q M X^p$. Let $f : \Lambda \to \Lambda$ be

$$f(P(X,Y)) = j(P(X - A, Y - B))$$

for $P(X,Y) \in \Lambda$. If $Q(X,Y)$ is a monomial corresponding to a divergent integral in the above sense, i.e., $Q(X,Y)$ starts with Y or ends with X, then we have $f(Q(X,Y)) = 0$. For any monomial $M \in \Lambda$, $f(M)$ is written as $M+N$ where N is a sum of monomials corresponding to divergent integrals. Since $f(N) = 0$ we have $f(f(M)) = f(M)$ for any monomial $M \in \Lambda$. Hence f is an idempotent endomorphism of Λ.

We set

$$\Gamma = 1 + \sum_{k \geq 1} \sum_{p_1,\cdots,q_k \geq 1} I_0(p_1, q_1, \cdots, p_k, q_k) X^{p_1} Y^{q_1} \cdots X^{p_k} Y^{q_k}.$$

Let us notice that the integral $I_0(p_1, q_1, \cdots, p_k, q_k)$ is convergent when $p_1, \cdots, q_k \geq 1$. Using the above Γ and $f : \Lambda \to \Lambda$, the Drinfel'd associator is expressed in terms of iterated integrals in the following way.

PROPOSITION 3.9. *The Drinfel'd associator* $\Phi(X, Y)$ *is expressed as*

$$\Phi(X, Y) = f(\Gamma)$$
$$= 1 + \sum_{k \geq 1} \sum_{p_1,\cdots,q_k \geq 1} I_0(p_1, q_1, \cdots, p_k, q_k) f(X^{p_1} Y^{q_1} \cdots X^{p_k} Y^{q_k}).$$

PROOF. By Lemma 3.8 we have

$$f(\Phi) = f\left(\lim_{a \to 0} a^{-Y} G^a (1-a) a^X\right).$$

Since $f(YM) = 0$ and $f(MX) = 0$ for any monomial $M \in \Lambda$ we obtain

$$f(\Phi) = f\left(\lim_{a \to 0} G^a(1-a)\right) = f(\Gamma).$$

Therefore, it is enough to verify that $f(\Phi) = \Phi$. For this purpose we consider the differential equation

$$G'(x) = \left(\frac{X-A}{x} + \frac{Y-B}{x-1}\right) G(x).$$

Using the solutions $G_1(x)$ and $G_2(x)$ of (3.9) we put

$$H_i(x) = x^{-A}(1-x)^{-B} G_i(x), \quad i = 1, 2.$$

We see that $H_i(x), i = 1, 2$, satisfy the above differential equation and that

$$\Phi(X - A, Y - B) = H_2^{-1} H_1 = \Phi(X, Y).$$

Hence we have $f(\Phi) = \Phi$. This completes the proof. $\square$

The above proposition together with Proposition 3.7 leads us to express $\Phi(X, Y)$ using multiple zeta values. Here is an expression up to degree 4 terms.

$$\begin{aligned}
\Phi(X,Y) = {} & 1 - \zeta(2)[X,Y] - \zeta(3)[X,[X,Y]] - \zeta(3)[Y,[X,Y]] \\
& - \zeta(4)[X,[X,[X,Y]]] - \zeta(4)[Y,[Y,[X,Y]]] \\
& - \zeta(1,3)[X,[Y,[X,Y]]] + \frac{1}{2}\zeta(2)^2[X,Y]^2 + \cdots.
\end{aligned}$$

We end this section by noticing that several relations among multiple zeta values can be obtained by considering the expression for knot invariants by iterated integrals. In [40], some linear relations among multiple zeta values have been obtained by comparing the equation (3.7) with the expression for $Z(K_0)$ using iterated integrals of ω_0 and ω_1. In particular, the relation

$$\left(\frac{1}{2^{2n-2}} - 1\right)\zeta(2n) - \zeta(1,2n-1) + \cdots + \zeta(\underbrace{1,\cdots,1}_{2n-2},2) = 0$$

was obtained.

3.2. Chern-Simons functionals and the Ray-Singer torsion

Let M be a compact oriented 3-manifold without boundary. We put $G = SU(2)$ and consider a principal G bundle P over M. The space of connections on P is denoted by $\mathcal{A}$. As explained in Section 2.5, $\mathcal{A}$ is identified with $\Omega^1(M,\mathfrak{g})$, the space of 1-forms on M with values in the Lie algebra $\mathfrak{g}$ of G. We recall that the Chern-Simons functional $CS : \mathcal{A} \to \mathbf{R}$ is defined by

$$CS(A) = \frac{1}{8\pi^2} \int_M \mathrm{Tr}\left(A \wedge dA + \frac{2}{3}A \wedge A \wedge A\right), \quad A \in \mathcal{A}.$$

The gauge group $\mathcal{G}$ is identified with the space of smooth maps $\mathrm{Map}(M,G)$ and the action of $\mathcal{G}$ is given by

$$g^*A = g^{-1}Ag + g^{-1}dg, \quad g \in \mathcal{G}.$$

The Chern-Simons functional transforms as $CS(g^*A) = CS(A) + n$, $n \in \mathbf{Z}$, under the action of the gauge group. The critical points of the Chern-Simons functional are the flat connections (Proposition 2.30).

Let us describe the infinitesimal action of the gauge group. Suppose that $g \in \mathcal{G}$ is written as $g = e^{tf}$ where f is a function on M with values in $\mathfrak{g}$. Then we have

$$g^*A = A + t(df + [A,f]) + O(t^2), \quad A \in \mathcal{A}.$$

Therefore, the infinitesimal version of the action of the gauge group is given by the covariant derivative d_A defined by

$$(3.10) \qquad d_A f = df + [A, f].$$

Let $\alpha \in \mathcal{A}$ be a flat connection and denote by $\mathfrak{g}_\alpha$ the local system associated with α. We denote by Ω_α^j the space of j-forms on M with values in $\mathfrak{g}_\alpha$. Then we have the de Rham complex

$$0 \to \Omega_\alpha^0 \to \Omega_\alpha^1 \to \Omega_\alpha^2 \to \Omega_\alpha^3 \to 0$$

with the differential d_α. Here $d_\alpha \circ d_\alpha = 0$ follows from the flatness of α.

In this section, we suppose that the flat connection α satisfies

$$H^*(M, \mathfrak{g}_\alpha) = 0.$$

By the Poincaré duality the above condition is satisfied if

$$H^0(M, \mathfrak{g}_\alpha) = 0, \quad H^1(M, \mathfrak{g}_\alpha) = 0.$$

The condition $H^0(M, \mathfrak{g}_\alpha) = 0$ signifies that the representation of the fundamental group $\pi_1(M)$ corresponding to the flat connection α is irreducible. The condition $H^1(M, \mathfrak{g}_\alpha) = 0$ implies that the connection α is isolated in the space of flat connections.

To motivate our study let us briefly recall the classical theory of asymptotic expansions. Our starting point is the Gauss integral

$$\int_{-\infty}^{\infty} e^{-\mu x^2} \, dx = \sqrt{\frac{\pi}{\mu}}, \quad \mu > 0.$$

By analytic continuation we obtain

$$(3.11) \qquad \int_{-\infty}^{\infty} e^{i\lambda x^2} \, dx = \sqrt{\frac{\pi}{|\lambda|}} \, e^{\frac{\pi i}{4} \frac{\lambda}{|\lambda|}}, \quad \lambda \in \mathbf{R}, \ \lambda \neq 0.$$

Let Q be a non-degenerate quadratic form in $x_1, \cdots, x_n$ and denote by sgn Q the signature of Q. By (3.11) we have

$$\int_{\mathbf{R}^n} e^{iQ(x_1, \cdots, x_n)} \, dx_1 \cdots dx_n = \frac{\pi^{n/2}}{\sqrt{|\det Q|}} \, e^{\frac{\pi i}{4} \operatorname{sgn} Q}.$$

For a real valued function $f(x_1, \cdots, x_n)$ we describe the asymptotics of the integral

$$(3.12) \qquad \int_{\mathbf{R}^n} e^{ikf(x_1, \cdots, x_n)} \, dx_1 \cdots dx_n$$

as $k \rightarrow \infty$. We suppose that the critical points of f are finitely many, non-degenerate and real. First, we deal with the integral in one variable

$$g(k) = \int_{-\infty}^{\infty} e^{ikf(x)} \, dx.$$

A basic principle of the *stationary phase method* is that the major contribution to the value of the above integral as $k \rightarrow \infty$ arises from a neighbourhood of a critical point of $f(x)$. For simplicity we suppose that $f(x)$ has only one critical point at $x = x_0$ and that $f''(x_0) > 0$. In this case we have

$$g(k) \sim \int_{x_0-\varepsilon}^{x_0+\varepsilon} e^{ikf(x)} \, dx = \int_{u_1}^{u_2} e^{ik(f(x_0)+u^2)} \frac{2u}{f'(x)} \, du$$

where we introduced a new variable u by $f(x) - f(x_0) = u^2$. By using

$$\lim_{x \rightarrow x_0} \frac{2u}{f'(x)} = \left(\frac{2}{f''(x_0)} \right)^{1/2}$$

we obtain

$$g(k) \sim \left(\frac{2}{f''(x_0)} \right)^{1/2} \int_{-\infty}^{\infty} e^{iku^2 + ikf(x_0)} \, dx$$

as $k \rightarrow \infty$. Combining with the Gauss integral (3.11) we obtain that the principal term of $g(k)$ as $k \rightarrow \infty$ is given as

$$\int_{-\infty}^{\infty} e^{ikf(x)} \, dx \sim \left(\frac{2\pi}{kf''(x_0)} \right)^{1/2} e^{ikf(x_0) + \frac{\pi i}{4}}.$$

In general, under the assumption for $f(x_1, \cdots, x_n)$ as above the principal term of the integral (3.12) is expressed as

$$\int_{\mathbf{R}^n} e^{ikf(x_1, \cdots, x_n)} \, dx_1 \cdots dx_n$$

$$\sim_{k \rightarrow \infty} \sum_{\alpha} \frac{\pi^{n/2} e^{ikf(\alpha)}}{k^{n/2} \sqrt{|\det \, H(f, \alpha)|}} \, e^{\frac{\pi i}{4} \, \text{sgn} \, H(f, \alpha)}$$

where the sum is for any critical point α of f and $H(f, \alpha)$ stands for the Hessian of f at α.

Let us go back to the situation of the Chern-Simons functional. We fix a flat connection $\alpha \in \mathcal{A}$ and denote by $(\Omega_\alpha^*, d_\alpha)$ the de Rham complex associated with α. Since the space of connections $\mathcal{A}$ is an affine space, its tangent space at α denoted by $T_\alpha \mathcal{A}$ is identified with the space of 1-forms $\Omega^1(M, \mathfrak{g})$. We fix a Riemannian metric on M. The tangent space $T_\alpha \mathcal{A}$ is equipped with an inner product defined by

$$\langle A, B \rangle = -\frac{1}{8\pi^2} \int_M \mathrm{Tr}\,(A \wedge *B).$$

We have an orthogonal decomposition with respect to this inner product

$$T_\alpha \mathcal{A} = \mathrm{Im}\, d_\alpha \oplus \mathrm{Ker}\, d_\alpha^*.$$

Here $d_\alpha : \Omega_\alpha^0 \to \Omega_\alpha^1$ is the covariant derivative defined in (3.10) and $d_\alpha^* : \Omega_\alpha^1 \to \Omega_\alpha^0$ is its adjoint operator. The subspace $\mathrm{Im}\, d_\alpha$ of the tangent space $T_\alpha \mathcal{A}$ is considered as the tangent space of the gauge orbit $\mathcal{G}\alpha$ at α.

Let us now compute the Hessian of the Chern-Simons functional at the flat connection α. For a one-parameter family of connections $\alpha_t = \alpha + t\beta$ with a parameter t we have

$$CS(\alpha_t) = CS(\alpha) + \frac{t^2}{8\pi^2} \int_M \mathrm{Tr}\,(\beta \wedge d_\alpha \beta) + O(t^3).$$

Therefore the Hessian at the flat connection α is the quadratic form

$$Q(\beta) = \langle \beta, *d_\alpha \beta \rangle.$$

Since the Chern-Simons functional is invariant under the infinitesimal gauge transformations, Q degenerates on the tangent space of the gauge orbit $\mathcal{G}\alpha$ and defines a non-degenerate quadratic form on the quotient space $\Omega_\alpha^1 / d_\alpha \Omega_\alpha^0$. The quadratic form Q on the above quotient space is identified with $- * d_\alpha$. We are now in a position to compute the determinant of this quadratic form. This is an infinite dimensional determinant and a precise meaning of it is the regularized determinant by the spectral zeta function as will be explained in the following. First, we consider the operator

$$P = \varepsilon(d_\alpha * + * d_\alpha)$$

acting on the space of differential forms

$$\Omega_\alpha^1 \oplus \Omega_\alpha^3 \cong \Omega_\alpha^1 \oplus \Omega_\alpha^0.$$

Here ε acts as 1 on Ω_α^0 and as -1 on Ω_α^1. The action of P on the direct sum

$$\Omega_\alpha^0 \oplus \mathrm{Im}\, d_\alpha \oplus \mathrm{Ker}\, d_\alpha^*$$

is expressed as

$$(3.13) \qquad P = \begin{pmatrix} 0 & -d_\alpha^* & 0 \\ -d_\alpha & 0 & 0 \\ 0 & 0 & Q \end{pmatrix}$$

and we have

$$(3.14) \qquad P^2 = \Delta_\alpha^0 \oplus \Delta_\alpha^1.$$

Here Δ_α^j is the Laplace operator on Ω_α^j defined by $d_\alpha^* d_\alpha + d_\alpha d_\alpha^*$ for $j = 0, 1$.

In general, the *regularized determinant* of the Laplace operator Δ is defined in the following way. Let

$$\lambda_1 \leq \lambda_2 \leq \cdots \leq \lambda_j \leq \cdots$$

be the positive eigenvalues of Δ counted with multiplicities. We define the *spectral zeta function* by

$$\zeta_\Delta(s) = \sum_{j=1}^{\infty} \frac{1}{\lambda_j^s}.$$

It is known that $\zeta_\Delta(s)$ is a holomorphic function when the real part of s is sufficiently large and is analytically continued to a meromorphic function on the complex plane (see [50]). In particular, $s = 0$ is not a pole of $\zeta_\Delta(s)$ and we have definite values $\zeta_\Delta(0)$ and $\zeta_\Delta'(0)$. The regularized determinant of Δ is defined by

$$\det \Delta = \exp(-\zeta_\Delta'(0)).$$

Let us go back to the flat connection α. It was shown by Ray and Singer that

$$T_\alpha(M) = \frac{(\det \Delta_\alpha^0)^{3/2}}{(\det \Delta_\alpha^1)^{1/2}}$$

is a topological invariant and does not depend on the choice of a Riemannian metric for M. We call $T_\alpha(M)$ the *Ray-Singer torsion* of M with respect to the flat connection α. Using the regularized determinant for Q^2 one can define $|\det Q|$. We have the following proposition.

PROPOSITION 3.10. *The regularized determinant of the Hessian of the Chern-Simons functional at the flat connection α is related to the Ray-Singer torsion by*

$$\frac{\sqrt{\det d_\alpha^* d_\alpha}}{\sqrt{|\det Q|}} = T_\alpha^{1/2}.$$

PROOF. By the expression of the operator P^2 in (3.14) we have the equality

$$\det P^2 = \det \Delta_\alpha^0 \, \det \Delta_\alpha^1$$

as the regularized determinant. This permits us to define $|\det P|$. Using the equation (3.14) we obtain

$$|\det P| = \det(d_\alpha^* d_\alpha)|\det Q|.$$

Thus we have

$$|\det Q| = \frac{(\det \Delta_\alpha^1)^{1/2}}{(\det \Delta_\alpha^0)^{1/2}}.$$

Combining with $\Delta_\alpha^0 = d_\alpha^* d_\alpha$, we obtain the desired equality. This completes the proof.   □

We end this section by mentioning expected asymptotic behaviour for Witten's invariant $Z_k(M)$ as $k \to \infty$, suggested from the formal expression as the Chern-Simons partition function

$$Z_k(M) = \int \exp(2\pi\sqrt{-1}k CS(A))\mathcal{D}A.$$

Let us recall that the critical points of the Chern-Simons functional are flat connections. The Chern-Simons functional is degenerated along the orbit of the gauge group and $\sqrt{\det d_\alpha^* d_\alpha}$ in the above proposition is interpreted as the ratio of the volume form of the gauge group and that of the Lie group. It is expected, as in a finite dimensional case, that the main contribution to the principal term of the asymptotic expansion of $Z_k(M)$ as $k \to \infty$ comes from the flat connections. It is known by the Cheeger-Müller theorem that the Ray-Singer torsion expressed by the regularized determinant of the Hessian coincides with a topological invariant called the Reidemeister torsion. The asymptotic behaviour of $Z_k(M)$ conjectured by Witten is

$$Z_k(M) \sim \frac{1}{2} e^{-3\pi i/4} \sum_\alpha \sqrt{T_\alpha(M)} e^{-2\pi i I_\alpha/4} e^{2\pi i(k+2)CS(A)}$$

where the right hand side is the sum for the flat connections α and I_α stands for the spectral flow defined by Atiyah, Patodi and Singer [8]. Let us recall that I_α is an integer determined modulo 8. A description of the asymptotic behaviour of $Z_k(M)$ involves a subtle problem concerning the framing of M and the phase factor. We refer the reader to Atiyah [6] for framings of 3-manifolds.

3.3. Chern-Simons perturbative invariants

We start with the case $G = U(1)$. Let $L = K_1 \cup K_2$ be an oriented framed link with two components in $\mathbf{R}^3$. We denote by P a principal $U(1)$ bundle over $\mathbf{R}^3$ and by $\mathcal{A}_{\mathbf{R}^3}$ the space of connections on P. The space $\mathcal{A}_{\mathbf{R}^3}$ is identified with the space of 1-forms on $\mathbf{R}^3$ since the Lie algebra of $U(1)$ is isomorphic to $\mathbf{R}$. In this case the Chern-Simons partition function corresponding to (2.28) is

$$Z_k = \int_{\mathcal{A}_{\mathbf{R}^3}} \exp \left(\frac{\sqrt{-1}}{4\pi} \int_{\mathbf{R}^3} A \wedge dA + \sqrt{-1} \int_{K_1} A + \sqrt{-1} \int_{K_2} A \right) \mathcal{D}A.$$

This is an abelian version of the Chern-Simons partition function. Let us notice that $A \mapsto A \wedge dA$ defines a quadratic form on the space of connections $\mathcal{A}_{\mathbf{R}^3}$. The original Chern-Simons functional for $SU(2)$ connections may be interpreted as an extension of this quadratic form to the non-abelian case such that it is invariant under the infinitesimal gauge transformations. The expression Z_k involves Feynman's path integral on the space of connections. Although it is hard to justify such integrals in general, the integral Z_k is one of the fairly well understood examples since it can be reduced to Gauss integrals.

Let us first explain a finite dimensional analogy. For a non-degenerate quadratic form

$$Q(x_1, \cdots, x_n) = \frac{1}{2} \sum_{i,j} \lambda_{ij} x_i x_j$$

we consider the integral

$$\int_{\mathbf{R}^n} e^{\sqrt{-1}(Q(x_1, \cdots, x_n) + \sum_{j=1}^n \mu_j x_j)} \, dx_1 \cdots dx_n.$$

By completing the square, we can reduce the above integral to the integral of

$$e^{-\sqrt{-1} \frac{1}{2} \sum_{i,j} \lambda^{ij} \mu_i \mu_j}$$

up to a constant multiple. Here (λ^{ij}) is the inverse matrix of (λ_{ij}).

Let us go back to the situation of the integral Z_k. The inverse of the differential operator d is an integral operator

$$(\widehat{L}\varphi)(\mathbf{x}) = \int_{\mathbf{y} \in \mathbf{R}^3} L(\mathbf{x}, \mathbf{y}) \wedge \varphi(\mathbf{y})$$

where the integral kernel $L(\mathbf{x}, \mathbf{y})$ is the *Green form* defined in the following way. For $\mathbf{x} \in \mathbf{R}^3 \setminus \{\mathbf{0}\}$ we put

$$(3.15) \qquad \omega(\mathbf{x}) = \frac{1}{4\pi} \frac{x_1 dx_2 \wedge dx_3 + x_2 dx_3 \wedge dx_1 + x_3 dx_1 \wedge dx_2}{\|\mathbf{x}\|^3}.$$

The above differential form gives a volume form of S^2 normalized as

$$\int_{S^2} \omega = 1.$$

The Green form is a 2-form on $\mathrm{Conf}_2(\mathbf{R}^3)$ defined by

$$L(\mathbf{x}, \mathbf{y}) = \omega(\mathbf{x} - \mathbf{y}).$$

It turns out that the integral Z_k is expressed as

$$\exp\left(\frac{\sqrt{-1}\pi}{k} \sum_{i,j} I(K_i, K_j) \right)$$

where $I(K_i, K_j), 1 \leq i, j \leq 2$, is the Gauss formula for the linking number

$$I(K_i, K_j) = \int_{\mathbf{x} \in K_i, \mathbf{y} \in K_j} \omega(\mathbf{x} - \mathbf{y})$$

if $i \neq j$. In the case $i = j$

$$I(K_i, K_i) = \int_{\mathbf{x} \in K_i, \mathbf{y} \in K_i'} \omega(\mathbf{x} - \mathbf{y})$$

is the self-linking number where K_i' is a curve on the boundary of a tubular neighbourhood of K_i' corresponding to the framing of K_i'.

Our next object is to discuss the asymptotics of the partition function of the $SU(2)$ Chern-Simons theory as $k \to \infty$. As explained in Section 3.2, it is expected that the principal term of the asymptotic expansion of the partition function is the contribution of the flat connections. Here we assume that any flat connection α satisfies $H^*(M, \mathfrak{g}_\alpha) = 0$.

First, we explain in a finite dimensional case a method to express the asymptotic expansion of the integral

$$Z_k = \int_{\mathbf{R}^n} e^{\sqrt{-1}k f(x_1, \cdots, x_n)} \, dx_1 \cdots dx_n$$

by means of Feynman diagrams. We deal with the case when f is written as

$$f(x_1, \cdots, x_n) = Q(x_1, \cdots, x_n) + \sum_{ijk} \lambda_{ijk} x_i x_j x_k$$

where Q is a non-degenerate quadratic form. By a change of variables we have

$$Z_k = k^{-n/2} \int_{\mathbf{R}^n} e^{\sqrt{-1}Q(x_1, \cdots, x_n)}$$

$$\times \sum_{m=0}^{\infty} \frac{(\sqrt{-1})^m}{m! \, k^{m/2}} \left(\sum_{ijk} \lambda_{ijk} x_i x_j x_k \right)^m dx_1 \cdots dx_n.$$

To obtain the asymptotic expansion of Z_k as $k \to \infty$ we compute the integral

$$(3.16) \qquad \int_{\mathbf{R}^n} e^{\sqrt{-1}Q(x_1, \cdots, x_n)} \left(\sum_{ijk} \lambda_{ijk} x_i x_j x_k \right)^m dx_1 \cdots dx_n.$$

By introducing new variables $J_1, \cdots, J_n$, we can express the integral (3.16) as

$$\left[\left(\sum_{ijk} \lambda_{ijk} D_i D_j D_k \right)^m \int_{\mathbf{R}^n} e^{\sqrt{-1}\left(Q(x_1, \cdots, x_n) + \sum_k J_k x_k\right)} dx_1 \cdots dx_n \right]_{J=0}$$

where D_j is given by

$$D_j = \frac{1}{\sqrt{-1}} \frac{\partial}{\partial J_j}.$$

Now we complete the square and observe that the integral (3.16) is written as

$$(3.17) \qquad \left[\left(\sum_{ijk} \lambda_{ijk} D_i D_j D_k \right)^m e^{-\sqrt{-1}\frac{1}{2} \sum_{ij} \lambda^{ij} J_i J_j} \right]_{J=0}$$

up to a constant multiple. This is a polynomial in λ^{ij} and λ_{ijk}, which is equal to zero if m is odd. In the case $m = 2$, the expression (3.17) consists of two parts

$$\sum_{ijki'j'k'} \lambda_{ijk} \lambda_{i'j'k'} \lambda^{ii'} \lambda^{jj'} \lambda^{kk'}$$

FIGURE 3.6. Two loop Feynman diagrams

and
$$\sum_{ijki'j'k'} \lambda_{ijk}\lambda_{i'j'k'}\lambda^{ij}\lambda^{kk'}\lambda^{i'j'}.$$

We express the above two types of expressions by trivalent graphs shown in Figure 3.6. These graphs are called two loop *Feynman diagrams*. In general, in the case when m is even, an m loop Feynman diagram is a trivalent graph with m vertices. The expression (3.17) can be written as the sum corresponding to m loop Feynman diagrams for even m.

Let us try to apply the above method of Feynman diagrams formally to the partition function

$$Z_k(M) = \int \exp(2\pi\sqrt{-1}kCS(A))\mathcal{D}A.$$

One of the main difficult points is that the Chern-Simons functional degenerates along the gauge orbit. In fact, the Chern-Simons functional is invariant under the infinitesimal gauge transformations and takes constant values along the gauge orbit. Hence the Hessian at any flat connection degenerates.

We explain the situation again in a finite dimensional case. Let G be an l-dimensional Lie group acting freely on $\mathbf{R}^n$ as isometries. Suppose that $f(x_1, \cdots, x_n) = Q(x_1, \cdots, x_n) + \sum_{ijk} \lambda_{ijk}x_ix_jx_k$ is invariant under the action of G. We assume that $F : \mathbf{R}^n \to \mathbf{R}^l$ is a smooth function such that F has only one zero point on each G orbit. Take $\mathbf{x} \in \mathbf{R}^n$ with $F(\mathbf{x}) = \mathbf{0}$ and consider the G orbit $G\mathbf{x}$. Let $\varphi : G \to \mathbf{R}^l$ be $\varphi(g) = F(g\mathbf{x})$ and denote by $J(\mathbf{x})$ the Jacobian matrix of φ at the unit $e \in G$. The determinant $\det J(\mathbf{x})$ signifies the ratio of the volume form of the G orbit $G\mathbf{x}$ and that of the Lie group G. The integral

$$Z_k = \int_{\mathbf{R}^n} e^{\sqrt{-1}kf(x_1, \cdots, x_n)} \, dx_1 \cdots dx_n$$

involves the same integral along the gauge orbit infinitely many times. We are going to extract its essential part

$$(3.18) \qquad \int_{\mathbf{R}^n} e^{\sqrt{-1}kf(x_1,\cdots,x_n)} \delta(F(\mathbf{x})) \det J(\mathbf{x}) \, dx_1 \cdots dx_n$$

where δ stands for Dirac's δ distribution. Applying the Fourier transformation, we have

$$\delta(F(\mathbf{x})) = \frac{1}{(2\pi)^l} \int_{\mathbf{R}^l} e^{\sqrt{-1}\sum_j F_j(\mathbf{x})\phi_j} \, d\phi_1 \cdots d\phi_l$$

and the integral (3.18) is written as

$$\frac{1}{(2\pi)^l} \int_{\mathbf{R}^{n+l}} e^{\sqrt{-1}(kf(x_1,\cdots,x_n)+\sum_j F_j(\mathbf{x})\phi_j)}$$
$$\times \det J(\mathbf{x}) \, dx_1 \cdots dx_n d\phi_1 \cdots d\phi_l.$$

For the Jacobian $\det J(\mathbf{x})$ we introduce Grassmann variables

$$\{c_i\}, \{\bar{c}_i\}, \ 1 \le i \le l,$$

satisfying the commutation relations

$$c_i c_j + c_j c_i = 0, \quad \bar{c}_i \bar{c}_j + \bar{c}_j \bar{c}_i = 0, \quad c_i \bar{c}_j = \bar{c}_j c_i, \quad 1 \le i, j \le l.$$

The determinant $\det J$ is written up to a constant multiple as

$$\int e^{\sqrt{-1}\sum_{ij} \bar{c}_i J_{ij} c_j} d\bar{c}_1 \cdots d\bar{c}_l dc_1 \cdots dc_l.$$

In this way by introducing new variables $\phi_1, \cdots, \phi_l$ and Grassmann variables $\{c_i\}, \{\bar{c}_i\}, 1 \le i \le l$, we can transform the integral (3.18) into a form such that $\delta(F(\mathbf{x}))$ and $\det J(\mathbf{x})$ are included in the action term. We refer the reader to [**47**] for Grassmann variables.

The above description is an analogy in the finite dimensional case. In the case of the Chern-Simons functional, to avoid the problem of degeneracy along the gauge orbit, we restrict to the space of connections A satisfying $d_\alpha^* A = 0$. This is a procedure corresponding to the restriction to $F(\mathbf{x}) = \mathbf{0}$ in the above finite dimensional case and is called a *gauge fixing*. For this purpose we introduce fermionic fields $c, \bar{c}$ and bosonic fields ϕ.

Motivated by the above heuristic argument, Axelrod and Singer [**10**] constructed 3-manifold invariants expressed as the integrals of Green forms associated with Feynman diagrams. Let M be a closed oriented 3-manifold. We fix a flat connection α of a principal $SU(2)$ bundle over M and suppose that α satisfies $H^*(M, \mathfrak{g}_\alpha) = 0$.

We fix a Riemannian metric and consider the integral kernel of $d^*_\alpha(\Delta_\alpha)^{-1}$ denoted by

$$L \in \Omega^2(M \times M; \mathfrak{g} \otimes \mathfrak{g})$$

where Δ_α is the direct sum of the Laplace operators Δ^j_α, $0 \le j \le 3$. Let $\{I_a\}$, $a = 1, 2, 3$, be an orthonormal basis of the Lie algebra $\mathfrak{g} = \mathfrak{su}(2)$ and express L as

$$L = \sum_{a,b} L_{ab}(x, y) I_a \wedge I_b.$$

We denote by $d_{(\alpha,\alpha)}$ the covariant derivative of the de Rham complex with coefficients in the local system $\mathfrak{g}_\alpha \otimes \mathfrak{g}_\alpha$ over $M \times M$. The above L is a Green form and satisfies

(3.19) $$d_{(\alpha,\alpha)} L_{ab}(x, y) = -\delta_{ab}\delta(x, y)$$

where $\delta(x, y)$, $(x, y) \in M \times M$, is a current of degree 3 which represents the Poincaré dual of the diagonal set $\Delta \in M \times M$ and satisfies

$$\int_{M \times M} \delta(x, y)\psi(x, y) = \int_M \psi(x, x)$$

for a 3-form $\psi(x, y)$ on $M \times M$. We have

$$L_{ab}(x, y) = -L_{ba}(y, x).$$

We denote by $\{\gamma_{abc}\}$ the structure constants of the Lie algebra $\mathfrak{g}$. Corresponding to the two loop Feynman diagrams in Figure 3.6 we set

$$I_{\Gamma_1}(M, \alpha)$$
$$= -\frac{1}{8} \sum \int_{M \times M \backslash \Delta} \gamma_{a_1 b_1 c_1} \gamma_{a_2 b_2 c_2}$$
$$\times L_{a_1 c_1}(x_1, x_1) \wedge L_{a_2 c_2}(x_2, x_2) \wedge L_{b_1 b_2}(x_1, x_2),$$
$$I_{\Gamma_2}(M, \alpha)$$
$$= \frac{1}{12} \sum \int_{M \times M \backslash \Delta} \gamma_{a_1 b_1 c_1} \gamma_{a_2 b_2 c_2}$$
$$\times L_{a_1 a_2}(x_1, x_2) \wedge L_{c_1 c_2}(x_1, x_2) \wedge L_{b_1 b_2}(x_1, x_2).$$

To define Green forms such as $L_{a_1 c_1}(x_1, x_1)$ we need to remove the divergent part on the diagonal set. We refer the reader to [10] for a precise definition. We fix a framing s of M, which is a trivialization of the tangent bundle TM. We denote by $CS_{grav}(g, s)$ the gravitational Chern-Simons invariant, i.e., the value of the Chern-Simons functional at the Levi-Civita connection for the Riemannian metric g. Let us notice that the gravitational Chern-Simons invariant depends on the choice of a framing s. A theorem due to Axelrod and Singer [**10**] is formulated in the following way.

THEOREM 3.11 (Axelrod-Singer). *Let M be a closed oriented 3-manifold. Under the above assumption for an $SU(2)$ flat connection α*

$$I_{\Gamma_1}(M, \alpha) + I_{\Gamma_2}(M, \alpha) - \frac{1}{4\pi} CS_{grav}(g, s)$$

is a topological invariant of M and does not depend on the choice of a Riemannian metric for M.

In the following, we give a brief sketch of the proof of the above theorem. To show that the quantity in Theorem 3.11 does not depend on the choice of a metric we consider a one-parameter family of Riemannian metrics $\{g^t\}$. Corresponding to $\{g^t\}$, we have a one-parameter family of Green forms L^t. We denote by ∇ the differential operator with respect to t. Since $d_{(\alpha,\alpha)}L^t$ does not depend on t, we have

$$d_{(\alpha,\alpha)}\nabla L^t = 0.$$

Therefore, by the assumption $H^*(M, \mathfrak{g}_\alpha) = 0$, there exists a 1-form B_t on $M \times M$ with values in $\mathfrak{g} \otimes \mathfrak{g}$ satisfying

(3.20) $$\nabla L^t = d_{(\alpha,\alpha)}B^t.$$

We express B^t as

$$B^t = \sum B^t_{ab}(x, y) I_a \wedge I_b.$$

Next, we construct a compactification of $\mathrm{Conf}_2(M) = M \times M \backslash \Delta$ as a manifold with boundary in the following way. We consider the real monoidal transformation $\pi : \widehat{M \times M} \to M \times M$ along the diagonal set Δ. The inverse image of Δ is identified with the tangent sphere bundle SM over M. We see that $\widehat{M \times M}$ is a compact manifold with boundary SM. The set of interior points of $\widehat{M \times M}$ is identified with $\mathrm{Conf}_2(M)$. The manifold with boundary $\widehat{M \times M}$ is considered as a compactification $\overline{\mathrm{Conf}_2(M)}$ of $\mathrm{Conf}_2(M)$. Then I_{Γ_1} and I_{Γ_2} are regarded as the integrals over the compact manifold $\overline{\mathrm{Conf}_2(M)}$, which confirms the convergence of I_{Γ_1} and I_{Γ_2}. Let us now compute ∇I_{Γ_i}, $i = 1, 2$, where the integrals are defined for the Green form L^t with

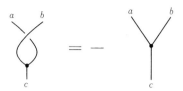

FIGURE 3.7. Anti-symmetry of the Lie bracket

a parameter t. By the Stokes theorem together with (3.19), (3.20), ∇I_{Γ_1} is expressed as the sum of the integral

$$\frac{1}{8}\int_{\overline{\mathrm{Conf}_2(M)}}\sum \gamma_{a_1b_1c_1}\gamma_{a_2b_2c_2}B^t_{a_1c_1}(x_1,x_1)L^t_{a_2c_2}(x_2,x_2)\delta_{b_1b_2}\delta(x_1,x_2)$$

$$+\frac{1}{8}\int_{\overline{\mathrm{Conf}_2(M)}}\sum \gamma_{a_1b_1c_1}\gamma_{a_2b_2c_2}L^t_{a_1c_1}(x_1,x_1)B^t_{a_2c_2}(x_2,x_2)\delta_{b_1b_2}\delta(x_1,x_2)$$

on $\overline{\mathrm{Conf}_2(M)}$ and an integral over $\partial\,\overline{\mathrm{Conf}_2(M)}$. The above sum of the integrals over $\overline{\mathrm{Conf}_2(M)}$ is written as

(3.21) $$\frac{1}{4}\int_M\sum \gamma_{acd}\gamma_{aef}B^t_{cd}(x,x)L^t_{ef}(x,x).$$

Similarly, from ∇I_{Γ_2}, we obtain the integral

(3.22) $$\frac{1}{4}\int_M\sum (\gamma_{ace}\gamma_{afd}-\gamma_{acf}\gamma_{aed})B^t_{cd}(x,x)L^t_{ef}(x,x)$$

over $\overline{\mathrm{Conf}_2(M)}$. In the equations (3.21) and (3.22) the sum is for any a,c,d,e,f. We apply the anti-symmetry of the Lie bracket and the Jacobi identity graphically shown in Figures 3.7 and 3.8 to show that the sum of (3.21) and (3.22) is equal to zero. To cancel the contribution from the integrals over $\partial\,\overline{\mathrm{Conf}_2(M)}$ we need the term CS_{grav}. We refer the reader to the original article [10] for more details.

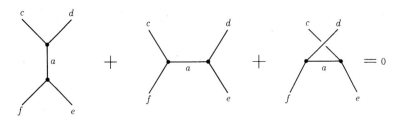

FIGURE 3.8. Jacobi identity

Further Developments and Prospects

Developments in conformal field theory

In this book we mainly treated conformal field theory of the Riemann sphere. A precise formulation of the space of conformal blocks by means of representations of affine Lie algebras was given for a Riemann surface of higher genus by Tsuchiya, Ueno and Yamada [52]. The space of conformal blocks is a finite dimensional vector space and forms a vector bundle over the moduli space of Riemann surfaces with a projectively flat connection. On a boundary point of the moduli space, where a Riemann surface is maximally degenerated as the union of spheres with three marked points, our connection is essentially the KZ connection. The projectively flat connection over the moduli space of Riemann surfaces is an extension of this KZ connection over the whole moduli space. In this way, conformal field theory is regarded as the theory of vector bundles over the moduli space of Riemann surfaces with projectively flat connections.

On the other hand, as explained at the end of Section 2.5, the space of conformal blocks is also formulated as the space of holomorphic sections of a complex line bundle over the moduli space of flat G bundles over a Riemann surface. In [9] a projectively flat connection on the conformal block bundle over the moduli space of Riemann surfaces was constructed from the viewpoint of the geometric quantization of the Chern-Simons gauge theory.

In this book Witten's invariant $Z_k(M)$ for 3-manifolds was formulated based on the holonomy of conformal field theory. It is an important problem to confirm, in the framework of conformal field theory, the asymptotic behaviour of $Z_k(M)$ suggested by the Chern-Simons gauge theory. Recently, the abelianization program of conformal field theory has been developed by Hitchin, Yoshida and several authors (see [26]). This approach will give rise to an explicit description of the holonomy of the projectively flat connection on the conformal block bundle in terms of theta functions.

In the case of the Riemann sphere, after the works [34], [51], Drinfel'd [18] established, in a general setting, a conceptual framework to describe the monodromy representations of the braid groups appearing as the holonomy of the KZ connection in terms of the R matrices in the theory of quantum groups. After this work, a deep study concerning a relation between conformal field theory and number theoretic aspects such as the absolute Galois group $\mathrm{Gal}(\overline{\mathbf{Q}}/\mathbf{Q})$ and the Grothendieck Teichmüller group has been pursued (see [18]).

Combinatorial aspects of finite type invariants

In the construction of Witten's invariants for 3-manifolds, we mainly focused on the geometric point of view based on conformal field theory, but Witten's invariants could be defined combinatorially based on representations of the quantum groups at roots of unity. A bridge between them is the fact that the holonomy of conformal field theory is described by means of the quantum groups at roots of unity. We refer the reader to [48] and [32] for the combinatorial definition of Witten's invariants.

Witten's invariant $Z_k(M)$ admits a symmetry by the Dynkin diagram automorphisms of the associated affine Lie algebra. In particular, in the case of $SU(2)$ invariants at level k with k odd, the cyclic group $\mathbf{Z}/2\mathbf{Z}$ acts on the set of weights $P_+(k)$ by $\mu \mapsto k - \mu$ freely and the invariant $Z_k(M)$ is decomposed into the product of a term described by abelian data such as the homology of M, and an essential part denoted by $\tau_r(M)$. Here we set $r = k + 2$. Let M be a homology 3-sphere. It can be shown that, if r is an odd prime, then $\tau_r(M) \in \mathbf{Z}[q], q = e^{2\pi i/r}$, holds. Expanding $\tau_r(M)$ in $q - 1$, and considering the situation where r is sufficiently large, Ohtsuki [44] constructed a topological invariant of M with values in the ring of formal power series

$$\tau(M) = \sum_{n=0}^{\infty} \lambda_n(M)(t - 1)^n.$$

According to a result due to Murakami [42], one has

$$\lambda_0(M) = 1,$$
$$\lambda_1(M) = 6\lambda(M)$$

where $\lambda(M)$ is the Casson invariant of M. A relationship between the above number theoretic expansion and the analytic expansion of $Z_k(M)$ as $k \to \infty$ has not been established and would be an important research problem.

The above $\lambda_n(M)$ is a typical example of invariants of finite type for 3-manifolds in the sense of Ohtsuki (see [45]). In [41], a universal invariant for such finite type invariants for 3-manifolds has been defined. The construction is based on Vassiliev-Kontsevich invariants for oriented framed links and Dehn surgery on such links. The above universal invariant is called the LMO invariant and takes values in the space of trivalent graphs modulo the relations shown in Figures 3.7 and 3.8 in Section 3.3. A relation between this universal invariant and $\tau(M)$ has been systematically studied in [45]. The Chern-Simons perturbative invariants discussed in Section 3.3 are determined for given flat connections over M, and take values in the same space of trivalent graphs. But a relation to the above LMO invariant has not been clarified.

As mentioned in the Introduction, the Jones polynomial was originally discovered in the framework of the theory of operator algebras. In the process of the study of the index of II_1 subfactors $N \subset M$, Jones obtained linear representations of the braid groups and the Markov trace. This was a starting point to define the Jones polynomial. The theory of the classification of II_1 subfactors has since been refined by Ocneanu and others and it turns out that the classification corresponds to that of combinatorial objects called paragroups. Here the fusion algebras and quantum $6j$ symbols play a role as in conformal field theory. Let us recall that quantum $6j$ symbols appear as connection matrices of 4-point functions in conformal field theory. For a 3-manifold we take a triangulation and associate to each tetrahedron a quantum $6j$ symbol. Turaev and Viro [53] defined a topological invariant of a 3-manifold by taking the state sum of the above quantum $6j$ symbols. This method is generalized to paragroups (see Evans and Kawahigashi [20]).

Topological invariants expressed as integrals

Chern-Simons perturbation theory gives rise to new integral representations for topological invariants and will shed new light on integral geometry. As a typical example, we mention a topological invariant for knots due to Guadagnini, Martellini and Mintchev [25] obtained in the study of Chern-Simons perturbation theory. Let ω

be a 2-form on $\mathbf{R}^3 \setminus \{0\}$ defined in (3.15). It is a Green form and determines a volume form of the sphere S^2 normalized as $\int_{S^2} \omega = 1$. Let K be a knot represented by an embedding $f : S^1 \to \mathbf{R}^3$. We fix a parameter for S^1 as $e^{2\pi it}$, $0 \leq t \leq 1$, and consider the integrals

$$I_X = \int_{0<t_1<t_2<t_3<t_4<1} \omega(f(t_1) - f(t_3)) \wedge \omega(f(t_2) - f(t_4)),$$

$$I_Y = \int_{0<t_1<t_2<t_3<1, \mathbf{x} \in \mathbf{R}^3 \setminus K} \omega(f(t_1) - \mathbf{x}) \wedge \omega(f(t_2) - \mathbf{x}) \wedge \omega(f(t_3) - \mathbf{x}).$$

The integrals I_X and I_Y define functions on the space of knots. It turns out that

$$I_X + I_Y - \frac{1}{24}$$

is a topological invariant of knots with values in integers and coincides with $a_2(K)$, the second coefficient of the Conway polynomial discussed in Section 3.1. This integral representation was obtained in [25] by a heuristic computation of Chern-Simons perturbative integrals. From a geometric point of view the above integrals I_X and I_Y are interpreted in the following way. We consider fibre bundles over the space of knots whose fibre is a certain configuration space of points so that the integrals I_X and I_Y are obtained by pulling back wedge products of Green forms and integrating them along the fibre. To show the topological invariance we need to prove $d(I_X + I_Y) = 0$ on the space of knots. Here $d(I_X)$ and $d(I_Y)$ are described by means of the compactification of configuration spaces. A principal tool is a graph complex whose boundary operator is given by the contractions of edges. We refer the reader to Bott and Taubes [15] for details. There are several subtle points for the higher order analogue of the above integrals, since one might have contributions from integrals on various strata of the compactified configuration space.

Consider an extremal situation when the knot K is close to its projection diagram and the volume form ω is concentrated on the north pole and south pole of the sphere. Then, as a limit, we obtain that I_X tends to a quantity determined only by crossing points of the projection diagram and that I_Y tends to an invariant of plane curves. This invariant of plane curves takes a constant value on each codimension one stratum of the space of plane curves. Such invariants have been studied systematically by Arnold [3] from the viewpoint of contact geometry. It will be interesting to investigate a further relationship between integral invariants coming from Chern-Simons perturbation theory and Arnold's invariants of plane curves.

In Section 3.1 we used the notion of chord diagrams for the study of Vassiliev invariants. From the point of view of Chern-Simons perturbation theory chord diagrams correspond to Feynman diagrams appearing in the perturbative expansion of the partition function of the Chern-Simons functional. Let Σ be a closed oriented surface. We define A_Σ to be the complex vector space spanned by the homotopy classes of continuous maps from chord diagrams to Σ modulo the 4 term relations. It was shown by Andersen, Mattes and Reshetikhin [1] that A_Σ has a structure of a Poisson algebra. This is a natural generalization of the Goldman bracket on the free homotopy classes of loops on Σ. For the Lie group $G = SU(2)$, we denote by $\mathcal{M}_\Sigma$ the moduli space of flat G connections over Σ. The moduli space $\mathcal{M}_\Sigma$ has a symplectic structure and the space of smooth functions on $\mathcal{M}_\Sigma$, denoted by $C^\infty(\mathcal{M}_\Sigma)$, is a Poisson algebra. We associate a representation of G to each loop of a chord diagram. Such chord diagrams are called G colored and we denote by A_Σ^G the space of G colored chord diagrams modulo the 4 term relations. Then one can construct a Poisson algebra homomorphism

$$\tau : A_\Sigma^G \to C^\infty(\mathcal{M}_\Sigma)$$

by adding information about the holonomy of a flat connection to the definition of the weight system in Section 3.1. It is expected that there is a universal Vassiliev invariant for links in $\Sigma \times I$ with values in A_Σ. This invariant would lead us to a deformation quantization of the Poisson algebra A_Σ which descends to $C^\infty(\mathcal{M}_\Sigma)$. This construction will give rise to a new point of view on the deformation quantization of $C^\infty(\mathcal{M}_\Sigma)$.

The space of conformal blocks may be regarded as a geometric quantization of the moduli space $\mathcal{M}_\Sigma$. It would be interesting to clarify a relation between the space of conformal blocks and the above deformation quantization of $C^\infty(\mathcal{M}_\Sigma)$. To establish Chern-Simons perturbation theory for 3-manifolds with boundary we will need to deal with the space of flat connections on closed oriented surfaces. The above method of chord diagrams on surfaces might be helpful in this direction. In the case $G = SL(2, \mathbf{R})$ the above τ for a loop γ gives the length of a minimal geodesic in the homotopy class of γ with respect to the hyperbolic structure associated with a flat G connection over Σ. Chern-Simons gauge theory for a non-compact group such as $G = SL(2, \mathbf{R})$ will be important for the purpose of understanding a relation to hyperbolic geometry. A conjecture due to Kashaev suggests that the asymptotic behaviour of certain quantum invariants detects the hyperbolic volume of the complement of a hyperbolic knot (see [30] and [43]). Recently, some important work has been done in this direction by several authors.

Relation to the moduli space of surfaces

First, let us recall matrix integrals. We denote by H_N the set of N by N Hermitian matrices $M = (M_{ij})$. For a polynomial $G(X)$ of a matrix X we consider the integral

$$\int_{H_N} \exp(-\text{Tr } G(M))dM$$

over the set of Hermitian matrices H_N. Here we regard H_N as a Euclidean space of dimension N^2 and dM is the invariant Lebesgue measure for H_N normalized so that

$$\int_{H_N} \exp\left(-\text{Tr}(M^2/2)\right) dM = 1.$$

The above matrix is relevant to the generating function for the number of ribbon graphs in the following way. Here a ribbon graph means a graph with a cyclic order for edges meeting at each vertex.

Consider the case

$$G(M) = \frac{M^2}{2} + \frac{tM^4}{4N}$$

as an example. We denote by $Z(N, t)$ the above matrix integral and compute the expansion of $Z(N, t)$ in t. In general, we put

$$\langle M_{i_1 j_1} \cdots M_{i_n j_n} \rangle = \int_{H_N} M_{i_1 j_1} \cdots M_{i_n j_n} \exp\left(-\text{Tr}(M^2/2)\right) dM$$

and compute it using the technique explained in Section 3.3. We obtain $\langle M_{ij} M_{kl} \rangle = \delta_{il} \delta_{jk}$, therefore, each term in the expansion of $\langle \text{Tr}(M^4)^m \rangle$ corresponds to a ribbon graph of valency 4 with m vertices. Attaching a disc to each connected component of the boundary of the ribbon graph, one can construct a closed oriented surface. We denote by $W(g, m)$ the number of ribbon graphs which give rise to a closed oriented surface of genus g. Considering the overlapping in the counting of graphs, we obtain

$$\log Z(N,t) = \sum_{g=0}^{\infty} N^{2-2g} \sum_{m=0}^{\infty} (-1)^m W(g,m) t^m.$$

By taking the dual cell decomposition of a ribbon graph on a surface, we see that $W(g,m)$ is also regarded as the number of ways of filling a closed oriented surface of genus g by m squares.

Let us denote by $M_{g,n}$ the moduli space of Riemann surfaces of genus g with n marked points. Each ribbon graph in the above construction corresponds to a cell in the cell decomposition of $M_{g,n} \times \mathbf{R}_+^n$ due to Mumford and Harer based on the theory of quadratic differentials. Here n is the number of discs attached to a ribbon graph. By specifying a metric for a ribbon graph, one can construct a complex structure on the associated closed oriented surface.

Around 1990, Witten [56] proposed the theory of 2-dimensional gravity and gave two ways to compute the n point function for its Lagrangian. The first one is the matrix integral related to the number of ribbon graphs on a Riemann surface and the second one is based on the intersection theory on the moduli space of Riemann surfaces. It was conjectured by Witten that the n point functions computed by the above two methods are identical. It suggests a relationship between the τ function of the KdV equation and the intersection theory on the moduli space of Riemann surfaces. The Witten conjecture was solved by Kontsevich [39]. On the other hand, in Chern-Simons perturbation theory, we observe that the two loop Feynman diagrams are ribbon graphs corresponding to the cells in the cell decomposition of $M_{0,3} \times \mathbf{R}_+^3$. The proof that the Chern-Simons perturbative integrals give rise to a topological invariant independent of a metric is based on the incidence relations of the ribbon graphs in the cell decomposition of the moduli space. This point of view would suggest a new stream of research involving topological field theory, integrable systems and Morse homotopy (see also Fukaya [23]).

Bibliography

[1] J. E. Andersen, J. Mattes, N. Reshetikhin, The Poisson structure on the moduli space of flat connections and chord diagrams, *Topology*, **35**(1996), 1069–1083.

[2] V. I. Arnold, The cohomology ring of the group of dyed braids, *Mat. Zametki*, **5**(1970), 227–231 (Russian).

[3] V. I. Arnold, *Topological Invariants of Plane Curves and Caustics*, University Lecture Series **5**, American Mathematical Society, 1994.

[4] E. Artin, Theorie der Zopfe, *Hamburg Abh.*, **4**(1925), 44–72.

[5] M. F. Atiyah, *The Geometry and Physics of Knots*, Cambridge University Press, 1990.

[6] M. F. Atiyah, On framings of 3-manifolds, *Topology*, **29**(1990), 1–8.

[7] M. F. Atiyah and R. Bott, The Yang-Mills equation over Riemann surfaces, *Philos. Trans. Roy. Soc. London Ser. A*, **308**(1982), 523–615.

[8] M. F. Atiyah, V. K. Patodi and I. M. Singer, Spectral geometry and Riemannian geometry I, *Math. Proc. Cambridge Philos. Soc.*, **77**(1975), 43–69; II *ibid.*, **78**(1975), 405–432; III *ibid.*, **79**(1976), 71–99.

[9] S. Axelrod, S. Della Pietra and E. Witten, Geometric quantization of Chern-Simons gauge theory, *J. Differential Geom.*, **33**(1991), 782–902.

[10] S. Axelrod and I. M. Singer, Chern-Simons perturbation theory, in *Proc. XXth International Conference on Differential Geometric Methods in Theoretical Physics, S. Catto and A. Rocha, eds.*, World Scientific, 1991, 3–45; II *J. Differential Geom.*, **39** (1994), 173–213.

[11] D. Bar-Natan, On Vassiliev knot invariants, *Topology*, **34**(1995), 423–472.

[12] A. Beauville and Y. Laszlo, Conformal blocks and generalized theta functions, *Comm. Math. Phys.*, **164** (1994), 385–419.

[13] A. A. Belavin, A. M. Polyakov and A. B. Zamolodchikov, Infinite conformal symmetry in two-dimensional quantum field theory, *Nuclear Phys.*, **B241** (1984), 333–380.

[14] J. Birman, *Braids, Links and Mapping Class Groups*, Princeton University Press, 1974.

[15] R. Bott and C. Taubes, On the self-linking of knots, *J. Math. Phys.*, **35**(1994), 5247–5287.

[16] K. T. Chen, Iterated path integrals, *Bull. Amer. Math. Soc.*, **83**(1977), 831–879.

[17] V. G. Drinfel'd, Quantum groups, in *Proceedings of the International Congress of Mathematicians, Berkeley, 1989*, 798–820.

[18] V. G. Drinfel'd, Quasi-Hopf algebras, *Leningrad Math. J.*, **1**(1990), 1419–1457.

[19] V. G. Drinfel'd, On quasitriangular Hopf algebras and a group closely connected with Gal($\overline{\mathbf{Q}}/\mathbf{Q}$), *Leningrad Math. J.*, **2**(1991), 829–860.

[20] D. Evans and Y. Kawahigashi, *Quantum Symmetries in Operator Algebras*, Oxford University Press, 1988.

[21] D. S. Freed, Classical Chern-Simons theory I, *Adv. Math.*, **113**(1995), 237–303.

[22] R. Fenn and C. Rourke, On Kirby's calculus of links, *Topology*, **18**(1979), 1–15.

[23] K. Fukaya, Morse homotopy and topological quantum field theory, *Comm. Math. Phys.*, **181**(1996), 37–90.

[24] K. Gawedzki, Wess-Zumino-Witten conformal field theory, in *Constructive Quantum Field Theory II*, G. Velo and A. S. Wightman, eds., Plenum Press, 1990, 89–120.

[25] E. Guadagnini, M. Martellini and M. Mintchev, Wilson lines in Chern-Simons theory and link invariants, *Nuclear Phys.*, **B330**(1990), 575–607.

[26] N. Hitchin, Flat connections and geometric quantization, *Comm. Math. Phys.*, **131**(1990), 347–380.

[27] M. Jimbo, A q-difference analogue of $U(\mathfrak{g})$ and the Yang-Baxter equation, *Lett. Math. Phys.*, **10**(1985), 63–69.

[28] V. F. R. Jones, Hecke algebra representations of braid groups and link polynomials, *Ann. Math.*, **126**(1987), 355–388.

[29] V. Kac, *Infinite Dimensional Lie Algebras*, Third Edition, Cambridge University Press, 1990.

[30] R. M. Kashaev, The hyperbolic volume of knots from quantum dilogarithm, *Lett. Math. Phys.*, **39**(1997), 269–275.

[31] C. Kassel, *Quantum Groups*, Springer-Verlag, 1995.

[32] R. Kirby and P. Melvin, The 3-manifold invariants of Witten and Reshetikhin-Turaev for $sl(2, \mathbf{C})$, *Invent. Math.*, **105**(1991), 473–546.

[33] V. G. Knizhnik and A. B. Zamolodchikov, Current algebra and Wess-Zumino models in two dimensions, *Nuclear Phys.*, **B247** (1984), 83–103.

[34] T. Kohno, Monodromy representations of braid groups and Yang-Baxter equations, *Ann. Inst. Fourier*, **37**(1987), 139–160.

[35] T. Kohno, Topological invariants for 3-manifolds using representations of mapping class groups I, *Topology*, **31**(1992), 203–230.

[36] T. Kohno, Vassiliev invariants and de Rham complex on the space of knots, in *Symplectic Geometry and Quantization*, *Contemp. Math.*, **179**, American Mathematical Society, 1994, 123–138.

[37] H. Konno, Geometry of loop groups and Wess-Zumino-Witten models, in *Symplectic Geometry and Quantization*, *Contemp. Math.*, **179**, American Mathematical Society, 1994, 139–160.

[38] M. Kontsevich, Vassiliev's knot invariants, in *I. M. Gel'fand Seminar*, *Adv. Soviet Math.*, **16**(1993), 137–150.

[39] M. Kontsevich, Intersection theory on the moduli space of curves and the matrix Airy function, *Comm. Math. Phys.*, **147**(1992), 1–23.

[40] T. Q. T. Le and J. Murakami, On Kontsevich's integral for the Homfly polynomial and relations of mixed Euler numbers, *Topology Appl.*, **62**(1995), 193–206.

[41] T. Q. T. Le, J. Murakami and T. Ohtsuki, On a universal perturbative invariant of 3-manifolds, *Topology*, **37**(1998), 539–574.

[42] H. Murakami, Quantum $SO(3)$-invariants dominate the $SU(2)$-invariant of Casson and Walker, *Math. Proc. Cambridge Philos. Soc.*, **117**(1995), 237–249.

[43] H. Murakami and J. Murakami, The colored Jones polynomials and the simplicial volume of a knot, math.GT/9905075.

[44] T. Ohtsuki, A polynomial invariant of rational homology 3-spheres, *Invent. Math.*, **123**(1996), 241–257.

[45] T. Ohtsuki, *Combinatorial Quantum Method in 3-Dimensional Topology*, MSJ Memoirs 3, Math. Soc. Japan, 1999.

[46] A. Pressley and G. Segal, *Loop Groups*, Oxford University Press, 1986.

[47] J. M. Rabin, Introduction to quantum field theory for mathematicians, in *Geometry and Quantum Field Theory, D. S. Freed and K. K. Uhlenbeck, eds.*, American Mathematical Society, Institute for Advanced Study, 1995, 185–269.

[48] N. Reshetikhin and V. G. Turaev, Invariants of 3-manifolds via link polynomials and quantum groups, *Invent. Math.*, **103**(1991), 547-597.

[49] D. Rolfsen, *Knots and Links*, Publish or Perish, Inc., 1976.

[50] R. T. Seeley, Complex powers of an elliptic operator, in *Singular Integrals, Proc. Sympos. Pure Math.*, **10**, American Mathematical Society, 1967, 288–307.

[51] A. Tsuchiya and Y. Kanie, Vertex operators in conformal field theory on $\mathbf{P}^1$ and monodromy representations of braid groups, *Adv. Stud. Pure Math.*, **16**(1988), 297–372.

[52] A. Tsuchiya, K. Ueno and Y. Yamada, Conformal field theory on universal family of stable curves with gauge symmetries, *Adv. Stud. Pure Math.*, **19**(1990), 459–566.

[53] V. Turaev and O. Viro, State sum invariants of 3-manifolds and quantum $6j$-symbols, *Topology*, **31**(1992), 865–902.

[54] V. A. Vassiliev, Cohomology of knot spaces, in *Theory of Singularities and its Applications*, American Mathematical Society, 1992.

[55] E. Witten, Quantum field theory and the Jones polynomial, *Comm. Math. Phys.*, **121**(1989), 351–399.

[56] E. Witten, Two-dimensional gravity and intersection theory on moduli space, *Surveys in Differential Geometry*, **1**(1991), 243-310.

[57] N. Woodhouse, *Geometric Quantization*, Second Edition, Oxford University Press, 1992.

Index

Titles in This Series

For a complete list of titles in this series, visit the
AMS Bookstore at **www.ams.org/bookstore/**.